THE COSMIC MATRIX

PIECE FOR A JIGSAW TWO

Anti-Gravity, Starships and
Unlimited Clean Free Energy

by
Leonard Cramp

The **New Science Series**:
- THE TIME TRAVEL HANDBOOK
- THE FREE ENERGY DEVICE HANDBOOK
- THE FANTASTIC INVENTIONS OF NIKOLA TESLA
- THE ANTI-GRAVITY HANDBOOK
- ANTI-GRAVITY & THE WORLD GRID
- ANTI-GRAVITY & THE UNIFIED FIELD
- ETHER TECHNOLOGY
- THE ENERGY GRID
- THE BRIDGE TO INFINITY
- THE HARMONIC CONQUEST OF SPACE
- VIMANA AIRCRAFT OF ANCIENT INDIA & ATLANTIS
- UFOS & ANTI-GRAVITY: Piece For a Jig-Saw
- THE COSMIC MATRIX: Piece For a Jig-Saw, Part II

The **Mystic Traveller Series**:
- IN SECRET TIBET by Theodore Illion (1937)
- DARKNESS OVER TIBET by Theodore Illion (1938)
- IN SECRET MONGOLIA by Henning Haslund (1934)
- MEN AND GODS IN MONGOLIA by Henning Haslund (1935)
- MYSTERY CITIES OF THE MAYA by Thomas Gann (1925)
- THE MYSTERY OF EASTER ISLAND by Katherine Routledge (1919)
- SECRET CITIES OF OLD SOUTH AMERICA by Harold Wilkins (1952)

The **Lost Cities Series**:
- LOST CITIES OF ATLANTIS, ANCIENT EUROPE
  & THE MEDITERRANEAN
- LOST CITIES OF NORTH & CENTRAL AMERICA
- LOST CITIES & ANCIENT MYSTERIES OF SOUTH AMERICA
- LOST CITIES OF ANCIENT LEMURIA & THE PACIFIC
- LOST CITIES & ANCIENT MYSTERIES OF AFRICA & ARABIA
- LOST CITIES OF CHINA, CENTRAL ASIA & INDIA

The **Atlantis Reprint Series**:
- THE HISTORY OF ATLANTIS by Lewis Spence (1926)
- ATLANTIS IN SPAIN by Elena Whishaw (1929)
- RIDDLE OF THE PACIFIC by John MacMillan Brown (1924)
- THE SHADOW OF ATLANTIS by Col. A. Braghine (1940)
- ATLANTIS MOTHER OF EMPIRES by R. Stacy-Judd (1939)
- SECRET CITIES OF OLD SOUTH AMERICA by H.T. Wilkins (1952)

THE COSMIC MATRIX

PIECE FOR A JIGSAW TWO

The Cosmic Matrix
Piece For a Jigsaw Part Two

Copyright 1999
by Leonard G. Cramp

First Printing

July 1999

ISBN 0-932813-64-X

Printed in the United States of America

Published by
Adventures Unlimited Press
One Adventure Place
Kempton, Illinois 60946   USA
auphq@frontiernet.net

10 9 8 7 6 5 4 3

THE COSMIC MATRIX

PIECE FOR A JIGSAW TWO

Anti-Gravity, Starships and
Unlimited Clean Free Energy

by
Leonard Cramp

THE SYNCOMAT MOTIVATED ALL PURPOSE VEHICLE

The family transport of the future. It will require no fuel, generate no pollution, noise, dust or radiation and its realisation is much closer than most people would believe.

THE COSMIC MATRIX

PIECE FOR A JIGSAW TWO
Anti-gravity, Starships and Unlimited Clean Free Energy

There is a certain logicality in physics which suggests that any new extension to cosmological theory is more likely to be correct if it satisfactorily helps to explain one or more existing apparent anomalies. This was not a prime intention of the author, but the development of this work certainly turned out that way. The reader will find that embodied within the main direction of the book there appears no less than six hitherto isolated areas of present day enquiry which will persuasively be seen to be totally reconciled by students of the representative studies. They are:

1. Astrophysicists may not necessarily agree, but may be intrigued to consider that the phenomena of gravity, magnetic and electrokinetics have indeed a common cause. Moreover so-called attraction and repulsion are one and the same thing. That the Big Bang and Steady State theories—together with the galactic red shift—are also completely reconciled.

2. Environmentalists will—among other things—come face to face with the realisation of a pollution free world.

3. Geologists will find the fascinating prospect of limitless colossal so-called free energy.

4. Space-flight engineers are presented with the possibility of a true vehicular space drive technique in which stress, strain and weight will cease to be a hindrance to motion.

5. Parapsychologists will at last find their subject is becoming of age and perhaps for the first time they will begin to see a unity in it all.

6. Ufologists not least—and perhaps of much more importance—it is the author's wish to share the coming well-deserved respectability of Ufology; after all that is where it probably all began.

All of the associated observations and scenarios in this book concerning paranormal and nano-engineering etc., were included in the original manuscript, which due to its potentially 'hot potato' content was tentatively shelved in 1987 by a well-known UK publisher, long before the recent upsurge of popular interest in these subjects.

Also it has been recently suggested that sprinkled throughout the text there are intentionally disguised informative clues. To date this has neither been admitted nor denied by the author.

The work includes the introduction of a theory which illustrates the existence of a gigantic universal carrier-belt transport system.

Works by the author
> Space, Gravity and the Flying Saucer. 1954, T. Werner Laurie.
> Piece for a Jigsaw. 1966, Somerton Publishing. 1996 Adventures
> Unlimited Press.

Leonard Cramp has a life Vice Presidency of BUFORA

CONTENTS

ACKNOWLEDGEMENTS

PART ONE
INTRODUCTION   11
PROLOGUE   29

PART TWO
1.   A WOOD FOR THE TREES SCENARIO?   37
2.   SETTING THE STAGE. Time, Pollution and Deadline 2000   65
3.   GRAVITY. The Ether, 'the Strong Force,' Matter and Energy   83
4.   TOWARDS THE UNIFICATION OF GRAVITY
     AND PRIMARY ENERGY   109
5.   THE COSMIC MATRIX. Was There Really a Big Bang?   121

PART THREE
INCLUSIVE ATTRIBUTES OF THE COSMIC MATRIX
6.   PARANORMAL ATTRIBUTES of the 'Spooky' Kind   155
7.   PARANORMAL ATTRIBUTES of the 'Nuts and Bolts' Kind   173
8.   ELECTROGRAVITIC MOTIVATION
     and Associated Phenomena, Including Crop Circles   217
9.   ELECTROMAGNETIC PROPULSION
     from Cars to Spaceships and Supermen   261
EPILOGUE: Not Least Unlimited Clean Free Energy   297
APPENDIX 1. Suppressed Inventions of the Corroborative Kind   323
APPENDIX 2. A Time for Everything   337
AUTHOR'S PROFESSIONAL BACKGROUND   355
BIBLIOGRAPHY   356
INDEX   357

ACKNOWLEDGEMENTS

Thirty years ago when my book *Piece for a Jigsaw* was published I began my acknowledgements by mentioning the acute sense of inadequacy I felt not only with regard to my portrayal of the subject matter in the book, but my attempt to express my gratitude to all those who had so kindly assisted me. Today although *The Cosmic Matrix* is the natural extension to that earlier work and due to the fact that with the passing of time, working mainly alone, I have lost touch with some of these helpers, while yet others have sadly long since departed, it seems appropriate for me to once again remember them all.

During the ensuing years much of my time has been taken up by many other projects, so the work on the subject of this book has of necessity been sporadic and closely shared by a very few members of my own family and to them I now formally extend my oft-repeated grateful thanks. In particular my wife Irene who *really* does know there is truth in the timeworn cliche about "the world's worst writer"; regretfully I sometimes feel that was invented by me. Not only has she endured years and become dexterous at deciphering my writing, but nowadays on occasions I have to ask her to decipher some of it *for me!* But the endless typing hours are only a part of this one lady's saga, in which she often politely endures my theoretical meanderings. Sometimes I am sure she hasn't understood a word I have said! Amazingly, often as not, it has been on such occasions when I have received my moments of eureka! Not surprisingly therefore that our twin daughters, Jane and Sue, should be apt to take all this for granted and continue to help a grateful father with their patience, belief and forbearance.

Although they would be the first to shun credit, I shall always consider myself to have been most fortunate to have the help of two very talented sons, David and Gary, who have always been there when Dad needed a helping hand. Repairing the workshop, house, garden or car one day, to intricate excursions in engineering science and test rigs the next, together with long debates concerning wiring diagrams and theoretical physics etc.; *that* is help unabating. But I am sure my words of thanks will best be focussed by success. Although to my knowledge I haven't been in touch with the one who calls himself Antony Avenel, I once again welcome this opportunity to offer my deep feeling of gratitude. I have long since ceased to ask the somewhat juvenile question of yesteryear, "why me?"

My constant gratitude is extended to George and Eve Bainbridge, who despite trials and tribulations have always shown much kindness and understanding towards me. To Peter Cross, one of the old bunch, whose timely encouragement and support helped immeasurably in making this book happen. Fiona for her much appreciated enthusiasm and hours spent photocopying. Peter Thorne, another stalwart of that elite bunch who unhesitatingly helped out with the artwork. His extraordinary talent and patient response to some of the more obtuse aspects of the work have never ceased to reassure me. To Colin Arnold, leading stressman and keen Ufologist, I am indebted for his continued help and kind recommendations for the publication of this work. To Timothy Good for his patience, belief and constant support.

I am also grateful to my lifetime friend Ron Howse who many years ago introduced me to the art of mirror grinding for his large telescopes. Perhaps in no better way is his patience reflected than by the occasion when having helped him to silver a mirror my tie inadvertently dragged a little on the polished surface, thereby ruining many hours of work. But Ron merely smiled and politely said, "Not to worry, it won't make much difference!" A brilliant design engineer who has accompanied me on this otherwise lonely journey, I shall always be grateful to Ron, in particular for his permission to quote the extraordinary case cited in Chapter 7.

When I first met John Batchelor in 1958 he was a very competent technical illustrator who often paid me compliments on some of my space pictures. Today he is internationally recognised as one of the world's leading technical artists, reflected by a workload which few would willingly bear. Despite this however he has continued to find time to motivate me and urge me to publish this work. It was with the generous assistance of our mutual friend David Meaning, designer and printer, that I finally took the first few tentative steps to do so. I have always been moved by the kind encouragement of the late Earl of Clancarty (Brinsley Le Poer Trench), when as a somewhat novice Ufologist I wrote my first book, *Space, Gravity and the Flying Saucer*. I like to think he would approve of this resulting effort.

Alan Hilton is not only one of the UK's most active and prominent Ufologists, he is also a kind and very helpful friend. His unhesitant and timely salvage of *Piece for a Jigsaw* from the pulp bin will always be a special memory for me. I would also like to convey my thanks and appreciation to David Haith for his spirited enthusiasm, for the copious copies of reports and recommendations and not least his optimism when the chips were down. And to Gary O'Neill for patiently helping out with the photography and other work in adverse conditions.

My special thanks are extended to David Hatcher Childress, prolific author and publisher who in the final analysis made this book possible. And to all those who may have directly or indirectly helped with this work.

PART ONE

INTRODUCTION

On a bright sunny day in May 1954, as a somewhat inexperienced unknown author, I had just arrived at the London office of publishers T. Werner Laurie to discuss the launch of my first book—*Space, Gravity and the Flying Saucer*—with the editor Waveney Girvan, when the office door suddenly burst open to admit a hurried and rather breathless gentleman who was subsequently introduced to me as Derek Dempster, Air Correspondent of the Sunday Express. Wondering what all the ensuing excitement was about, my already acute sense of inadequacy took on new depths when I learned that by special request Dempster had just delivered the first copy of the book to Buckingham Palace and a further copy was required!

Now for every first-time author acceptance of a book is gratifying, but to have a request for the first copy to be hand-delivered to Buckingham Palace was an unexpected bonus. Today, over four decades later, I can look back detachedly with some amusement and with the benefit of hindsight still vividly recall my dreamworld reaction as I stood there that bright sunny day. Then, everything seemed to be so clear-cut and uncomplicated. Now I reflect more soberly on the correctness of my decision to yield to another publisher's persuasion to offer my subsequent findings here, for excerpts from that first book were serialised in several American newspapers, being the third ever published on the subject and the first with a corroborative scientific appeal. Yet before the first batch of books could be sold, the British Book Publishing Centre of New York coincidentally and strangely went into voluntary liquidation. All six thousand copies of *Space, Gravity and the Flying Saucer* were withdrawn and pulped. A check on the situation a year later revealed the company was back in business *under the same name!*

At that time of course I was already acquainted with the worldwide bias concerning UFOs, for I had been rudely introduced to it on many occasions. Even so, still rather naive, as a British Interplanetary Society member I remember being utterly bewildered to learn that Waveney Girvan had been ignored in his request of Arthur C. Clarke—who had published his belief in the merits of the G. Field propulsion principle—to allow supportive quotation in this respect. At that time it was well known that Clarke had little sympathy for UFO reports. I was yet to learn that one with so much undoubted vision, could also be so blind.

For me these incidents were the forerunners of a series of events which were to be repeated and significantly influence and stimulate my private research into energy and space vehicle propulsion for the rest of my life.

I was to be encouraged by the kind interest shown by the Prime Minister Sir Winston Churchill and to meet with scientists, pilots and people from all walks of life who had been involved in UFO close encounters, some of whom, 'for the sake of their reputations', found it prudent to remain silent.

In those early times I could not have suspected that I would investigate many incidents with an engineer's unbiased scrutiny and enthusiasm and on occasion stand in silent wonder at the implications of what I saw and even less be involved with a UFO CE3[1] incident which would include no less than <u>eight</u> members of my family (coincidentally gathered together while on holiday at my home). Not least the frantic reaction of our dog Lassie whose barking heralded the beginning of an event of such unquestionable reality which to my knowledge has seldom been equalled to this day!

I was to meet with people of Intelligence rank who obviously knew far more about the domestic and political side of such issues than I, as a consequence of which I would later ask myself the question, "If I knew more about the reality of this subject than my local Member of Parliament—who happened to regard UFOs as nonsense and myself as little more than a tolerably harmless eccentric—and our government consists largely of people such as he, having not the slightest notion of this vitally important phenomenon, who was running the country?" Self-evidently *not* the government! Since those days I have been somewhat reconciled to this extraordinary state of affairs and have had my suspicions vindicated—if not suitably reinforced— by the more frivolous informative source known as the BBC's television series "Yes Prime Minister"!

I could not have known that my exertions would eventually lead me into little known paths and gain me access to even less accessible information—including correspondence between our astronauts—and I was to learn the truth in the tenet "Seek and ye shall find", so that despite popular current belief to the contrary, I now have good reasons to accept that not only is our galaxy teeming with other forms of intelligent life, but we are <u>not</u> alone in our solar system.

In 1966 I had the opportunity to publish my second book, *Piece for a Jigsaw.* This was a private venture largely determined by the need to avoid a repetitious fate of *Space, Gravity and the Flying Saucer*. Once again I tried to offer some credence for the existence of UFOs by pointing out that at least in terms of my proposition for Gravity Wave motivated spacecraft, many of the observed UFO behaviourisms—far from being at variance with Newtonian physics—did in fact make jolly good scientific sense. I filled the entire book explaining how such vehicles *could* indeed execute right-angled turns at high speed or come to an abrupt stop in mid-air, would be silent and almost certainly on occasions be dazzlingly luminous. In fact identical to the behaviourisms of UFOs. The statistical odds for such similarity being mere coincidence was of course quite untenable.

As an independent publishing venture which we could carefully monitor, we considered there was little that could go wrong, yet despite our efforts it did go unbelievably wrong! An account of the bizarre circumstances involved is unfortunately too long to relate and in any case would be out of place in this volume; suffice to add that the book was ruined, had to be pulped and reprinted. Consequently we lost the advantage of costly advertising. However numbers of the reprint were sold around the world, but due to a long stay in inadequate storage most of these also had to be disposed of. Recently *Piece for a Jigsaw* has been reprinted in America as a paperback and it is hoped that this may vindicate many of those early letters I received from appreciative readers kindly claiming the book to be "unique" and "offering as it does technical corroboration, it remains among the best UFO books to be written," etc., etc.

However this isn't just another UFO book, rather it is an account of what I have learned about the nature of space, energy and vehicular propulsion expanded by a close rapport with them. Moreover although some of this information is not mine to give, it is given freely, *it is true* and vitally important to everyone. It contains a message, it contains a friendly warning, but perhaps of more importance, to many it may offer hope.

In my constant searching I have been directed towards a means of using energy correctly which could rid us of the awful consequences of using raw nuclear processes irresponsibly. I have glimpsed exciting alternative methods of transportation which not least could take us to the stars. Any honest aerospace engineer would agree we are hardly likely to realistically achieve that by employing relatively crude chemical rockets. Therefore although the following presentation is my own, I cannot and do not accept total credit for it and I am content to leave any underlying origin to the intuitive sense of the discerning reader.

In taking the decision to offer this work for publication I am acutely aware of the need for it to reach as wide a readership as possible, but this is not easy due to the fact that the main topic is necessarily of a technical nature. However the basic ideas have been approached in analogous form in a semi-technical and more appealing manner for most readers of an enquiring nature to assimilate. Therefore it is emphasised that some of these ideas should be interpreted as representative rather than definitive.

In order to achieve this somewhat daunting task I have taken advantage of lessons learned as a result of many lectures I have given, in which perforce I often had to draw on extensive use of models and mechanical aids, primarily to set the stage and offer a quick stimulant to reassure and even help some members of the audience to keep awake! The short story in the prologue is no exception, though it is nonetheless qualitatively technically correct.

Although at first glance some of the introductory background material in Chapter 1 may appear somewhat diverse, it has nevertheless been considered important enough to acquaint the reader with it, if for no other reason than to help build a rapport and in anticipation of some questions which will be raised. From this it will be seen I have spent most of my working life in the Aerospace industry, but unsponsored research can be a big drain on one's income, therefore a dual occupation as designer/consultant to several world leading toy manufacturers had to be somehow fitted in.

Now if to some this may seem incongruous, it should be pointed out that modern day toy manufacture is in fact full-size engineering in miniature, calling for the same—in some instances even greater—high degree of sophistication. Therefore a stint in the toy manufacturing business offers excellent know-how experience for the design and construction of small bench top equipment for the laboratory. In retrospect this was most fortunate and the extraordinary thing about it was I didn't seek the toy trade, it sought me!

But of no less interest and relevance to this report, is the fact that competition in this trade is second to none, where secret research, surveillance and even sabotage is not unknown—as I know to my cost. Such experience has largely influenced my world view of the status quo where invention is concerned, which of course has a direct bearing on the decision to present my conclusions in the form of this book. By way of example, a typical event involving a highly sensitive industrial venture with which I was involved is carried in Chapter 1, which may help to answer some further anticipatory questions.

As with any radically different theory one can always expect some prejudice from within the scientific community and there can be little doubt the subject of this book will prove to be no exception. Therefore it is considered the example offering evidence of the damaging effects from this inertial tendency in Chapter 1 is not only warrantable but necessary in further support of the decision to publish these findings here. Moreover due to the main semi-technical format it is logical that some readers may appreciate the inclusion in Chapter 1 of the more generally presented background anecdotes to this one man's saga.

Such echoes from past experiences have prompted me to stress that from the beginning of this report there are several points I should make clear. The first being that although on occasion I have criticised some disproportionately orientated people among the higher echelons of scholarship for their apparent lack of foresight, equally I want to assure the reader that I am acutely aware of—and have diligently tried to avoid—the tendency toward 'inverted snobbery' which successful self-learning can invoke. From a very early age I went through this tortuous 'school' driven by an insatiable desire to go fishing for answers among natural phenomena and what answers I came up with were due solely to stubborn persistence more than being a good angler.

Not having access to many books and libraries, I often dabbled in exciting experiments which to me were pet discoveries, only to be disillusioned later to learn that I had unknowingly been repeating work which learned gentlemen (including Michael Faraday) had made famous many years before! However this boyhood intuitive faculty was not restricted to the world of nuts and bolts, but extended to the domain of space, stars and planets and in particular strange paranormal effects I experienced. Even so, despite the well-meaning intentions of some of my peers when they tried to reassure me that I have been fortunate in not being penalised by the blinkers of a more formal education, I—like many others—have consequently suffered a degree of in-built inferiority which the lack of formal education in the sciences can bring.

It is natural and not uncommon for some readers to want to learn something of an author's background and in particular the influence of the formative years. So I have responded to the suggestion that something appropriate might be of interest.

I was born in London in a typical poor family, being the second youngest son of four children. When he was nine years of age, my elder brother's health was failing—due in part to the London smog of those days—and accordingly he was sent to convalesce in an old country estate health centre in the south of England. Unfortunately he was pining for his family and the idea proved counterproductive, so it was suggested that I should be sent to join him.

But at seven years of age I was one of the youngest kids there and I found the long dark dormitory very scary, not helped by the sinister rumbling noise created by the wind-driven well pump supplying water to an overhead tank situated in the clock tower above my head!

The windows of our dormitory ran along one complete wall with our beds head to tail facing them. I shall never forget one particular sultry night when I was awakened by hushed murmurings from a group of the lads who—bathed in 'moonlight'—were looking out of the windows. Apparently they had spotted something unusual for I can remember one of them saying "course it's not a star, it's moving." By now very much awake, I was horrified to see the young lad—whose bedhead was immediately opposite the foot of my bed—gracefully ascending! With head thrown back, as if looking up at the ceiling, and arms outstretched he alighted on my bed—as I thought sleep-walking—saying "Len, Len, look at that beautiful lady." My response was to promptly dive under the blankets!

Of course this is only history for me now, but with the benefit of hindsight I have since realised that I can trace the beginning of my fascination in aerospace matters back to that time. Also, although my schooling suffered badly as a result of my stay in Ongar, my ability as an artist knew no bounds. So much so, that when I eventually returned to my old London school, on one memorable occasion I remember being asked to stay late one evening until my parents arrived, given paper and pencils, put into an otherwise empty classroom and told to draw! In my innocence I wondered what the hell all the fuss was about. But I can also remember visiting the local corner shop to scrounge empty cardboard boxes on which I could sketch boats and airships and anything that could fly. There was the day when my alarmed mother saw me hop off the roof of my father's shed, presumably in the hope that the family umbrella would act as a parachute! And even later, the time when I was preparing a coal-gas filled party balloon complete with a net crocheted by my sister which was to support a small basket to take the family kitten! Still, the evening session at the school did provide a bonus, for I was given all the paper and pencils I required. After that the empty cardboard boxes had an extended use—making models!

One summer evening in 1939 found me having tea with my sister and her friend Gwen, and Irene—the lovely young girl who was to become my wife. At that time—having never really recovered from my lost schooling—academically I was a total dunce. However after a series of extraordinary jobs—including as a 'sugar boiler' at Tate and Lyle's (which cost me a badly scalded hand) I eventually found a niche doing the thing I could do best, drawing, as a commercial artist.

That summer evening in 1939, my sister's friend offered to read the tea leaves in our cups; apparently she had a propensity for this art. Of course I can't remember much of what was said, except something she told me that my wife also distinctly remembers. She said there would be two women in my life! This has been perfectly true, for the past thirty years we have cared for my sister. But bearing in mind my complete ineptitude about technical matters at that time, plus the date—when the advent of UFOs hadn't even begun—the next projection was astonishing. For looking puzzled, Gwen had said, "You know I can't understand what this is all about, but all I can get with you is engines, engines everywhere. but strange kinds of engines, they're like round shapes and they are spiralling all around your head!" Bearing in mind my acute sensitivity about my lack of ability in the three Rs, I was frankly amused when Gwen next said, "I'm sorry, I don't know what it means, but I can say it appears to be very important and one day you will know and be well known concerning it."

Perhaps the reader will understand when I admit that as the years have rolled by I have occasionally been humbled and have recognised a strange empathy with all self-opinionated sceptics, educated or otherwise. For in those days, when I could barely read, write or spell, how could anyone have convinced me that today I would be writing this introduction to a book in which, among other things, I would be offering what could be the beginning of a new-age transport/energy technology? How two of the major cosmological theories of our time—Sir Fred Hoyle's "Steady State" and the "Big Bang"—might be shown to be aspects of one and the same thing, and thus be reconciled, as shown in Chapter 5. How the age-old apparently different phenomena of attraction and repulsion are also one and the same, as shown in Chapter 8. The answer has to be—just like the sceptics—neither Gwen nor anyone else could possibly have persuaded me to accept such a preposterous idea. Yet just as Gwen forecast, eventually much of it all has become dispassionately clear to me, and today Irene and I are very proud to share its significance with Gary and David, our two exceptionally talented sons who nowadays—all too often it begins to seem to me—are light years ahead.

By the way, that evening Gwen also told us that we would be 'comfortable,' but never very rich. So considering the near poverty we were enduring at that time…? Not least she was the first—though by no means the last—to advise Irene and I that one day we would be blessed with twins: two lovely daughters, Jane and Sue, for whom to this day we are equally grateful and justifiably very proud.

_ As will be seen from my book *Piece for a Jigsaw Three* and Appendix 2 herein this somewhat trying excursion has not been without rewards along the way, yet they tend to pale into insignificance in the light of the main subject of this book. Moreover I welcome this opportunity to offer some encouragement to similarly disadvantaged youngsters of today by pointing out that persistently applied searching eventually brings results, as many have found in all walks of life. In my own case, from being a backward young kid who could do little more than draw, the process eventually earned me senior positions in two highly secret special projects departments in Britain's leading aerospace companies. The account in Chapter 2 is usefully supportive of the foregoing and in addition it helps to illustrate the present 'wood from the trees' situation with which mankind is probably faced.

Although as a youngster I was always fascinated by the phenomena of gravity, I also had an innate acceptance of the *normality* of some aspects of the so-called paranormal. Therefore in later life I was excited to witness evidence of it. But what puzzled me was the obvious truth, that among the admitted sceptics few *wanted* to be convinced anyway! I continued to be puzzled when I met such people from the scientific community and felt instinctively they were missing something which otherwise could be enormously important to them. Somehow I felt they were like people looking at a painting too closely, so that they couldn't observe the whole.

The first 'breakthrough' in this respect occurred when I met Eric S., an extremely brilliant young industrial physicist. At the beginning of our friendship he too was very sceptical about the paranormal and I would ask how could he be so vehemently certain about the subject without even taking the trouble to study it? Eric was one of the few who took that important step back. Today he can persuasively discuss aspects of UFOs and the paranormal as a veteran!

In 1953 I was extremely fortunate to be introduced to the 'Unity of Creation' theory presented in Chapter 4. This helped me immeasurably insofar as I now had a common basis with which to correlate otherwise seemingly unrelated phenomena—including gravity *and* the paranormal. Therefore, whereas scientists the world over had tried desperately to research gravitational effects but were penalised by its exceeding weakness, I realised that a form of anti-gravitational effects had been going on so-called paranormally since the beginning of time under the guise of 'mystic' levitation. How utterly remarkable—though perhaps significant—that physicists knew so little about it! Today more and more people <u>are</u> beginning to accept phenomena of the kind discussed in Chapters 6 and 7. Even so, despite the 'non-repeatability in the laboratory' claim, there can be little wonder that those relative few who *did* believe and studied years ago, should now have the advantage of a head start. I have little doubt that before long it will be taken for granted that the so-called *paranormal* will take its rightful place as being just as *normal* as any other facet of natural phenomena. Moreover, I suggest we shouldn't be surprised if this includes time and teleportation, as is shown in Chapters 2 and 7.

Although such reconciliation would seem to be a relatively short step to take and developing the necessary technological hardware in terms of space flight and all vehicular propulsion (typified in Chapters 5, 8 and 9 together with the summation of what might well prove to be a broader unified field theory) may be an exciting predictable prospect, there can be little doubt that given the validity of the hypothesis, it is the possibility of a new inexhaustible 'free' energy acquisition which is at present of paramount importance.

Any meaningful appreciation of the repercussions on a society which the adoption of a radically different transport/energy system could have can only be gained when viewed within the context of the associated social issues, such as economic, political, environmental, etc., but if it is at all necessary, only a passing acquaintance with some available statistical information—such as that included in Chapter 2—is sufficient to persuasively convey the grim alternative. The behind the scenes establishment inertia and intrigue are all part of the general scenario in this Pandora's box and only by including some of these pieces—quoted in Chapters 1 and 2—can the stark truth concerning our questionable future be revealed. Some will not like it very much, it is not pleasant to be shown how juvenile and irresponsible some of us can be. Sadly many will not even care, while those who may find it more comfortable might regard the subject of this book as little more than pure conjecture on the author's part. In which case they would be well advised to adopt an open mind, for given the incentive I suggest it might prove to be an accurate forecast of the kind of energy and transport technology we could have in the not too distant future. But of no less importance is the fact that it is a dream which cannot realistically be *the sole prerogative of any one faction or any one nation on earth!* In furtherance of this work it is my hope I may merely be pointing the way.

There are reasons to suspect—some of which are discussed in Chapters 1 and 2—that work involving the broad principle outlined in this book has been—and almost certainly still is—going on all over the world. As long ago as the 1950s there was news of research into so-called 'antigravitics' being conducted by several countries, including India, Russia and America. Indeed as shown in Chapter 1 this author was quite innocently involved in a somewhat hilarious demonstration of evidence for it! Therefore I think it can be taken for granted that among the higher portals of, say, NASA and the American government, for example, there still exists clandestine operations. Having worked in the aerospace industry for most of my life, I am acutely aware of the need for secrecy, but unfortunately in the UK this is frequently carried to an unbelievably ludicrous extreme.

During the past few years it has become an almost common experience to meet with younger members of the community who have a very keen interest in the subject of alternative energy. One hears of more and more alternative energy and 'new age' societies being formed whose membership ranges from the very lay to a few professional scientists and the author can quite easily recognise the same kind of enthusiastic fervour with which UFO and astronomical societies began to spring up all over the world during the 1960s. Without wishing to offend it seems as though once a subject has made its mark and become popular, almost anyone wants to "get in on the act."

As discussed in Chapter 1, no such fervour or clamour existed when as a young man I first witnessed paranormal levitation. Indeed society at large hadn't even begun to know that rockets weren't anything other than toy fireworks, when by invitation I became a member of the embryonic British Interplanetary Society. Therefore my interests have never been inspired by such public persuasion. Indeed without any sense of bigotry I think it fair and correct to say that I have gained the advantage of many years working alone, initiated not by fashion of the times, but by a keen desire to understand, which bore fruit and earned me the right to be encouraged by an all-embracing theory of the cosmos on which I could continue my search. In addition to which I think it no less fair to point out that few—if any—of the more recent claims are founded on such a theoretically oriented basis. Moreover neither do any of these even mention an inherently acquired vehicular propulsion principle as being part of an integral claim. Indeed, as stated it was solely due to initially trying to understand gravity that I began to suspect that space and matter are inextricably one and the same. Therefore a solution to gravitational manipulation by implication should reflect a tenable energy source.

At this juncture it may be fittingly correct for me to yield to the opinion of readers concerning a certain amount of confusion which has bedevilled some of my earlier work.

While it is a fact that for some reason both *Space, Gravity and the Flying Saucer* and *Piece for a Jigsaw* were successfully 'nobbled,' in their own way both received quite a degree of notoriety. As stated, *Space, Gravity* was serialised in several American newspapers and *Piece for a Jigsaw* has been described by Timothy Good[2] as "a classic among UFO books." Both books were the first to offer a technical treatment of the subject and both were liberally cited in *UFOs: An Annotated Bibliography* by the USAF Office of Scientific and Aerospace Research, edited by Lynn E. Catoe (Library of Congress 1969). So despite their early curtailment there was, and still is, adequate published information for researchers concerning them. How strange it seems therefore that although quite a few subsequent authors have not only described many of the technical observations I made—such as thrust vectoring to tilt, repulsion field effects, gravity field effects, differential G effects between crew and vehicle, absence of aerodynamic drag, heating and noise, isolation from acceleration forces, complete with supportive diagrams etc.—none of them give even a hint of these pioneering efforts. Indeed sadly the very latest and by far the best among this bunch—despite a lengthy reference list of no less than *one hundred and thirty-four* works and names cited—reference to my work still remains conspicuous by its absence! Beyond a degree of natural apathy towards such a trend among society, there can be little doubt that in the final analysis such total disrespect contributed towards the decision to offer my eventual conclusions in this book. It is worth adding, despite the semi-technical nature of *Piece for a Jigsaw*, I was grateful in having succeeded in my efforts to convey some more obtuse aspects of the subject to a fairly wide spectrum of people and of those who did find the going a little tough, they responded to the recommendation that they 'skip the sums' but give careful consideration to the conclusions. I suggest the credibility they kindly afforded me then will not be amiss with *The Cosmic Matrix/*

During the last three decades or so there have been references to mystical black boxes which were supposedly capable of achieving some extraordinary things. In the 1950s for example, there was the black box of 'Radionics' coined by its inventor, George de la Warr. At that time—and to the best of my knowledge even to this day—the extraordinary images reproduced on normally unexposed photographic plates was never disproved. On the contrary, during subsequent court proceedings—on which occasion the term 'black box' was coined by the media—the judge ruled de la Warr's case as satisfactorily established. Unfortunately de la Warr died not long after, but his workers—including physicists from the contemporary establishment—still continue to conduct research into 'radionics.'

Nowadays it is common to hear designers over the drawing board using the colloquial version, David's or John's black box position, and in more sombre vein, black box is the press jargon name for the in-flight recorder fitted to most large aircraft, and not least, it is a black box carried by every American president's military aide wherever he goes, which could make the difference of peace on Earth or Armageddon!

Then there is the small domestic box—popular with some early science fiction writers—who assumed that one day in the future people will strap on to their homes and be provided with all the private power they need for heating, lighting and cooking etc.

Strangely the pattern of fate casts familiar shadows and it is hoped that having read this book, perhaps for the very first time, the reader will be able to examine a new concept which appropriately enough could be labelled as a box, a *box of prodigious power and awe-inspiring potential.*

Interestingly, in addition to the more obvious examples above, the story of Pandora's box could be something more than just a quaint simile to our present energy 'crisis' situation and in this context it may be appropriate to review the essence of the story from Greek mythology.

Of PANDORA'S BOX, the concise Oxford Dictionary says:

The box in which Hope alone remained when by its rash opening all objects of desire were dispersed to play havoc among mankind.

And of PANDORA, the Longman's Modern English Dictionary says:

The first woman, sent to earth as a punishment for Prometheus's crime of *stealing fire from the Gods,* Zeus gave her a box which, when opened, let loose all human misfortunes, *but Hope remained in the bottom to comfort mankind.*

Such an interpretation may be forgiven, for we <u>have</u> stolen and <u>grossly</u> dissipated the world's resources, and few of us can deny that it <u>has</u> helped to cause 'havoc among mankind' and—as we shall see later—there <u>is</u> hope remaining in the bottom of the box.

In furtherance of this simile the author would like to be regarded as just another individual who has had the good fortune to peep into the box as the lid was prised up a little further and who would like to share some of the fascinating contents he has spied. Having said that I feel in good company, for as I pen these words I am fully mindful of the fact that with a high degree of probability, somewhere at some time the reader will have shared such an experience, and therefore will be able to accept my apologies for the somewhat autobiographical tone which of necessity accompanies part of the text.

Based on empirical work suggested by theoretical considerations involving considerable research over many years, the conclusions are those of the author and it must be reiterated, at this time it is prudent o advise that they be regarded as merely indicative and that which is offered here is for all those with responsible motivations.

In fairness to those who have encouraged me it is necessary to bring attention to the fact that among the many people I have befriended over the years there are regretfully some of their associates in various parts of the world today—some holding positions of prominence—who have on several embarrassing occasions been quick to adopt personal identity with this work, when in truth they weren't even born when (as shown elsewhere) some of the author's efforts were published and I gave public lectures on the subject all too many years ago. Though the balance has often been redressed by the many letters of support I have received, including some from young scientists of the present whose initial interest was kindled by the Unity of Creation theory, an interest and sympathy which they hold to this day. Not least the reader's attention is drawn to the apparently precognitive tendency of other works shown in the Appendix. I have very important reasons for asking that similar precognitive overtones shouldn't be excluded from the main directive of this book.[3]

On occasion, sometimes in the depth of a winter's night and a poorly heated workshop, an eavesdropper might have overheard my sons and I enviously speculating on how far ahead such and such an organisation were in this work. That 'they' were researching G effects was beyond question, that they had all the required resources was equally certain, but as to whether their research was purely random or theoretically based was anybody's guess. In the natural order of things, given the correct theoretical basis and supportive resources, there is no contest. But where are the definitive results? To which Ufologists of this age would reply "in establishments like Area 51." But I am not so sure about that, for although in the normal turn of events this would appear sound enough, reflection suggests that it is only likely to be true if other more generously equipped researchers are at least sympathetic towards certain paranormal phenomena, for—as will be shown in this book—they would be seriously disadvantaged without it.

My position is—and always has been—that there exists two interpretive extremes concerning an understanding of some natural phenomena, with unbelieving diehard scientists at one end of the spectrum and the totally inept—though nevertheless sincere—lay paranormal lobbyists at the other. For my part I consider myself fortunate in having in-depth experience at both ends, which has enabled me to nurture a more balanced view, though I admit to having suffered somewhat in this irksome divide! An oft-repeated example being that in which some of the advanced but nonetheless 'ordinary' aerodynamic VTOL concepts on which I have worked, are quite erroneously 'identified' by some Ufologists as weird unidentified spacecraft!

Perhaps in no better way can I quantify this situation than by remarking that only recently I was informed by Gordon Creighton (editor of Flying Saucer Review) that due to the nature of the material in my second book *Piece for a Jigsaw* in 1964, the majority of Ufologists had assumed me to be very much of "a nuts and bolts man." Whereas in the 1950s when *Space, Gravity and the Flying Saucer* was published the *scientific* lobbyists were inclined to regard me as a "psychic phenomena buff." In which case logically the more liberal stand I have adopted—despite the lack of massive funding—may in fact have provided very much of a head start in this research. Concerning this my conclusions may of course be somewhat at variance, but somewhere along the line I expect other approaches to this work will inevitably converge. Neither do I expect my proposed acronym 'Syncomat' will be adopted; time will tell. Also it should be added, it is not without due consideration that I have chosen to close this introduction in a similar vein with which it began, only in the following instance it wasn't a bright sunny day in London 1953, but a very dark night on Silpho Moor near Scarborough, North Yorkshire in November 1957. At that time a couple were startled by the arrival of a glowing mass from the sky. Extraordinary as it may seem the object was retrieved by the police and delivered to one Antony Avenel, who they believed to be a local 'UFO expert.' Sometime later Avenel wrote to me saying that although he no longer had the object he realised in hindsight it had been intended for me.

Then at a Rotarian meeting in Hampshire in May 1959, Lord Dowding, Commander-in-Chief of Fighter Command during the Battle of Britain, told fellow Rotarians that the object had been examined by him and several experts. Describing it he said, "It was a metallic disc of about three feet in diameter and appeared to be manufactured of extremely tough copper-like metal." After considerable efforts it was finally separated into two halves with the aid of oxyacetylene cutters. Lord Dowding claimed that to the surprise of everyone involved the internals of the object proved to be a soot-coated void containing nothing but a strange metallic book packed with strange hieroglyphics in a language unlike any found on Earth. Eventually these were deciphered by an expert, one Philip Longbottom, who having devoted considerable time to the task concluded:

The whole thing is not just a simple substitution code, but is a very complicated effort. To make up a complete 'language' like this would seem to be out of all proportion to a hoax, however elaborate. Like any other translator, one tends to get 'inside' the thoughts and feelings of the person who wrote the original, and I firmly believe that this is not a 'made-up' language, but one in constant use. The whole thing flows so easily, and yet contains the natural mistakes that one would expect, considering the difference between our written and spoken word.

On receiving the correspondence from Antony Avenel concerning this matter, I was naturally disappointed, but like so many other Ufologists and scientific peers of those days, I in my questionable wisdom took solace from the fact that there was—literally it seemed—little in it, no semblance of a radio, no power plant etc., just a strange metallic book. However in the light of the material offered in the final chapters of this volume, together with the modern trend towards 'Nano Engineering'[4] .....the conclusions are best left to the reader.

An article concerning this incident duly appeared in a book titled *UFOs over Hampshire and the Isle of Wight* by Robert Price, who quoting from a Ministry of Defence official said:

Paragraph 2 of the Ministry's letter also states... 'we do not, nor do we have any need to, interview individuals who might have witnessed an unusual incident.'

> This statement may come as a surprise to Police Constable Len Haffenden, 'who spent almost three quarters of an hour on the telephone being closely questioned by a senior RAF intelligence officer who displayed more than a passing interest in his experience. Telephone interviews, it seems, do not count.'
> The letter ends in a somewhat caustic tone by insisting that the MOD is not in possession of any positive evidence that 'alien spacecraft' have landed on Earth—but has it already hidden away the greatest find in the history of Ufology? Have we already seen the arrival on our planet of an extra-terrestrial space probe, small and unmanned though it may have been?
> If this amazing claim is true, and judging by the undoubted integrity of the source, there is no reason to suppose that it is not, then where is this book today? Has it been made a subject of the Public Records Act and filed away for thirty years? If this is the case, can we now expect it to be released for public scrutiny in the near future?
> It seems fittingly correct to close this introduction by including the following last passage from *Piece for a Jigsaw* in which I said:
> There is, of course, far more to the flying saucer story than the mere mechanical aspect I have chosen to demonstrate; in truth the story only begins here. It had been my hope to talk about a personal experience of thought-provoking nature, of the Mother-ships, why I have not theorised about them in this text, more about the energy the discs use, the little signposts which indicate where those who want to, may look for it themselves. I wanted to offer my own findings on where the big carriers may

be coming from; how they come; nay, why they bother to come at all. I had planned to offer this for those who would care to listen at the end of this book; indeed, some of it has already been written. But in presenting the technically corroborative evidence for the flying saucers, I have found too much to say, too little space to say it in, and thinking about it now, I feel it right and proper to let it stand for what it is, so that my carefully planned contents sheet has had to be cut right through the middle. Reflecting on it with some regret, I am partially reconciled into thinking that whereas what has been offered in these pages may at least make a few more people think, what was to come after would almost certainly have put some of them off. For, to some extent, it represents a different order. Its proper place must be between separate covers. In making this decision, I realise that many folks who might have borne the semi-technical nature of this work in order to read what was to follow, will not now do so. To them I can only extend my apologies with my feeling of deep privilege, and renew my efforts to complete the half—their half—of a fascinating story.

With the benefit of hindsight it is now abundantly clear that what I had to say then would certainly have put some readers off. Now, over three decades later, the situation has dramatically changed, I can talk the paranormal language with an ever-increasing number of physicists who are becoming aware of that other half of this fascinating story (Chapters 6, 7 and 9).

Finally it should be stressed the author's conclusions reached in this book are completely original—as evidenced by the authority of quite a few published works. They are based on many years of searching spread over a lifetime and of course it is highly possible they may be seriously at fault. However nowadays it is becoming a common occurrence to hear of gravity and energy 'breakthroughs'; remarks such as '180 degree phase shift' or 'pulsed field stimulation' are rather emptily bandied around, theoretical support remaining significantly absent. It is my fervent hope that such hot air talk, together with the tendency for missing laboratory documents, industrial sabotage, patent and industrial sharks—together with other instances quoted in the following pages, that my family and I have endured—may in some measure answer the oft-raised questions, "why not a technical paper" and "despite the restrictions imposed on two popular books, why a third?"

[1] Close encounter of the third kind (UFO and occupants) described in my book, *The AT Factor, The UFO Connection,* to be published as a collection of works by the author.

[2] The top UFO expert and author of several best-selling books including *Beyond Top Secret: Alien Liaison* and more recently *Alien Base.*

[3] Liberally illustrated in *The AT Factor.*

[4] Molecular-sized engineering.

PROLOGUE

You are up and about early this fine spring morning in anticipation of taking the new family car for an acquaintance spin. You traded in the Mk.1GRM (ground reference mode) Hummingbird series model a week ago for the sleek X15 Falcon newly delivered yesterday and now residing in your garage.

For the past few years, the Hummingbird has been a good old friend and you still feel a pinch of regret parting with it. Not that there could be any comparison, for GRM Hummingbirds are after all little more than surface monitored wagons with a maximum speed of 80 mph at a monitored ground height of 10 feet over rough terrain, or 200 mph at 30 feet over water, EM controlled arterial roads and motorways. Not that you could have done either before decentralisation of industry.

For all that, they were pleasant enough little tubs fitted with the updated silent Delta drive units. In the early days, some of the lads had got round to tweaking the tuning a bit, which was after all a simple wiring job. But dashing around all over the place, at low altitude and high speed—without a sound in the dead of night—was not your brand of Russian roulette and it *did* rather annoy the neighbours! Even so, as a youngster, you and your pals had a lot of fun cannibalising some of the old-time piston driven cars and fitting a set of wheels to redundant early Hummingbird body shells. Remembering some of the bruising incurred, you smilingly muse, "There's a lot to be said for EM suspension!"

As you approach the garage, you reflect that living in the country has its advantages, for you are not restricted to GRM operation in order to get out of town as 'townies' are. Even so, a short trip down the road will help blow the cobwebs away until you get the feel of the new bird.

The clang of the up and over door leaves you standing in rapt admiration for your new runabout, you liked the look of the 2 plus 2 seat version right from the first glimpse in the brochure; compact, racy and just right for Beryl and the two kids. The 'stardust' colour job looks great too, for which the brochure claims a 25% drag reduction at 10,000 feet. She's smaller than the equivalent old conventional 'wheelies,' due of course not only to the long-redundant, bulky heavy suspension units of that era, but also the propulsion/lift units being not much bigger than medium-sized vacuum flasks. "Looks more like a bird with stowed wings," you muse, "probably had a bearing on the tag of the series."

Garaged and standing on the short static pads, the Falcon is normally just over four feet high, but this morning, hovering statically and silently in GRM during the night, it reaches almost to the top of your head. It provides a more comfortable ingress and the power consumed will not even show on the meter, not that you are concerned, for your last year's total supply from the new power pack you had installed, amounted to less than last week's milk bill! There is, of course, the six months' maintenance charge, but there is so little in these things that you are quite capable of overhauling it yourself.

Switching off the power supply and unplugging the cable, you make a mental note of the fact that you now have two weeks' maximum operational margin before the next recharge, but, due to the on-board, fail-safe automatic auxiliary recharging system, there's little to be concerned about. A close pass of your signet ring and the canopy rides up silently to reveal the spacious interior, sheer luxury of the upholstery and automatic contouring seats. Almost like a greeting, the little craft dips barely perceptibly as you slide into the seat, close the canopy and check the automatic seat belt.

Falcon's on-board programmed computer, now registering owner/operator acquisition, gives all systems green status as the console display lights confirm auxiliary power supply, duration, etc. Aware as you are of the system's basic fail-safe simplicity, you are additionally reassured by the quadrupled redundancy capability. The armrest side-stick, which has long since replaced the old-fashioned steering wheel, responds to your touch, and Falcon eases silently forward down the lawn-bordered drive on to the roadway. For about one and a half miles, the beautiful little machine glides smoothly along, with only a slight murmur of the parting air, until you reach the convenient clearing you have in mind. After a brief stop, you take Falcon up to hover at 20 feet and initiate the in-flight mode command, deploying wings and control surfaces; your little runabout has now become an extremely efficient high speed aircraft. Having satisfied yourself on the feel of the controls, you settle back and programme for lift-off.

Although you are quite used to it by now, the thrill of rapid vertical transition ascent never ceases to impress you, and you pull away into an almost silent, breathtaking upward turn back towards your home. At 500 feet you bring Falcon to a static hover, enjoying the view of the house and garden. "The garden," you muse, "must trim that back lawn." Beryl and the children are up and about by now, and are outside waving at the family's new pride and joy. Just for the hell of it—or perhaps a little showing off—you turn Falcon onto its back, hanging there rotating like some up-turned large flying turtle! With the old helicopters in mind, you cannot resist thinking "try that one for size," as you recover the little craft to a more dignified posture. A waggle of wings and Falcon is in a streaking vertical climb towards the sky.

Falcon as a sophisticated car

*Falcon as a sophisticated VTOL aircraft with
optional flying surfaces*

Aerodynamically, the little craft responds beautifully, the indicated air speed is now 280 miles per hour—*upwards!* Not only does the thrust unit automatically pro-vide matching aerodynamic thrust vectoring control, but it doesn't require a medium in which to function, therefore it will continue to propel the machine vertically at this speed into airless space; in a little while your bird will have become a spacecraft, and in terms of the early space age reckoning, you will have become an astronaut. Having obtained a satellite traffic-free zone and getting aerospace clearance, you settle back to enjoy the rest of the trip.

With ever-decreasing drag, the little machine hurtles towards the void beyond, and in just under half an hour you are in space some 150 miles above the earth. You reduce speed to zero, level off, and select static hover mode. The same thrust (equals weight) power which sustained you in your playful gambolling a short time ago, sustains you now in the serene beauty of airless space. Should you require it, the designers and builders of the Falcon have used hard-earned know-how from the old rocket launcher days to provide you with full space environmental systems for several days. During the whole time you would experience none of the weightlessness that the early astronauts endured, for unlike them you are not in orbit.

You spend nearly half an hour gazing spellbound, and taking photos of the beautiful planet beneath. Then a glance at the dash-clock reminds you to look up and to the west expectantly. For a minute there is nothing but the star-studded backdrop to see, then a barely discernible movement of one of them catches your eye as it gets perceptibly larger. Within a few moments the sleek lines of a Mark 2 Aerovan come into sight. It is piloted by your next-door neighbour, Bob Rutland, who, with his family, has been out on a weekend trip. To meet up with them was an additional reason for you to take Falcon for a short hop. Bob is a good friend, and had helped arrange the purchase of your runabout. He has three children: two girls and a boy, and like Doreen, his wife, they are thoroughly devoted touring types. Bob eases the Aerovan closer so they can all give you a friendly wave.

About 48 feet in diameter of lenticular form with spaced-out sets of windows, the Aerovan sleeps up to six comfortably. It is completely devoid of aerodynamic control surfaces; the older Mark 1 model did have a couple of fins, but the Mark 2 is completely IMVC (impulse motor vectored controlled). You smile appreciatively and reflect, "so what price flying saucers?" You and Bob chat over the radio for a few minutes before they descend homeward. "They had a great time by the sound of it," you muse.

Highly pleased with your own new acquisition, you point Falcon's nose almost vertically down towards the earth at a controlled, comfortable rate of descent. On re-entering the atmosphere, you experience no buffeting or aerodynamic heating as the old rocket vehicles used to, for you have selected a maximum air speed of 290 knots, though had you so desired, you could have descended at an air speed of anything between that and zero.

On the upward leg of your trip, Falcon acquired the rotational speed of the earth, but with increasing height you would have lagged behind a little, however the acquired displacement has been compensated for by the on-board computer. So on breaking cloud base you have no difficulty in finding your way back home. There is now no sign of the Rutlands.

During the next few minutes, you occasionally check the return route readouts while considering the possibility of a trip to your favourite part of the coast next week. There's a very nice stretch of beach you know of, which you and the family frequent, with good offshore underwater diving facilities, just the spot to check out the Falcon's sub-aqua performance. At this you cannot help reflecting; not so very long ago, it would have seemed impossible for a machine to operate in the air or space, or even under the sea, without a propeller or jet of some kind. "Ah well, that's progress." You have just spotted the familiar terrain and another five minutes brings you hovering over the small local clearing at about 30 feet. Feeling rather elated and more than a little 'over the moon' with Falcon, you initiate GRM and check the metamorphosis of the flying control surfaces retract-and-lock. You have been away a little over one and three quarter hours!

Bob and the family live quite near here, but you resist the temptation to await their arrival and set Falcon down to a couple of feet GRM. A few minutes later, and you are stepping out onto your front lawn. The children come out to greet you as you give Falcon—now hovering silently—an appreciative pat. You leave her there, as Beryl will want you to take her and the kids shopping later. She's out back somewhere. "Oh no, that bloody lawn."

Almost any sci-fi writer could deal far more ably with the above imaginary scenario. It is not an attempt to portray what *may* happen in the far distant future, but based on the implementation of the technology presented in the remainder of this book, what *could* happen in an ideal world in the early years of 2000.

The live-in Aerovan VTOL aero-spacecraft.
The general arrangement drawings for these vehicles have been
included in the author's book 'The AT Factor'

PART TWO

1

A WOOD FOR THE TREES SCENARIO?

Strange Gifts from Strangers

During the last few months of 1953 and the early part of 1954—while the preparation of the manuscript for *Space, Gravity and the Flying Saucer* was under-way—some extraordinary events coincidentally occurred, three of which, in hindsight, might be regarded as significant.

First, a technical author friend of mine drew attention to a fascinating article (published in the then-popular *Practical Mechanics*) titled "The Unity of Creation Theory"[1] by Antony Avenel (pseudonym), who incidentally received the small UFO disc in 1957 mentioned in the introduction.

At that time, I had nurtured a somewhat mechanistic interpretation of gravity as being a kind of etheric flow which could be expressed in purely hydrodynamic terms. As this was to be included in the manuscript—and there seemed to be a degree of visual parity with the unity of creation theory—my friend suggested that mention of the latter independent work might be supportive. I agreed this made sense and accordingly wrote to Mr. Avenel through the magazine, seeking his opinion.

Not only did he respond immediately, but he offered me the right to use the entire article if I so wished. Naturally, I was delighted to accept this and during the next year or so we corresponded regularly, during which time I was always deeply impressed by his continued kindness, patience, and generosity, of a degree which most people would find extraordinary.

Accordingly, the Unity of Creation theory was included in both my books, *Space, Gravity and the Flying Saucer* in 1954 and *Piece for a Jigsaw* in 1964 and now repeated again in this present volume. Suffice to add that it was the first published theory which described space with a *structured* form, yet it has never been publicly acknowledged nor recognised; the reader may well ponder why? Since then, other writers—particularly in the USA—some of amateur status as well as professional, have adopted the general idea without identifying its origin, using more exotic terminology such as 'space juice,' etc., while some have even proffered a density value (typically of the order 10^{32}grams/cm^3). Avenel ventured a wavelength for his 'creative rays' as less than 10^{-13}cm.

The second event with which I was involved occurred in February 1954, when two boys, Stephen Darbishire and Adrian Myer, took the—now well-known—Coniston UFO photographs. When the story hit local headlines, the same friend who had drawn my attention to Avenel's article—and who incidentally was a UFO sceptic—pointed out that I had the know-how to carry out some qualitative analysis between the Coniston photo and those taken by George Adamski, insofar as the outline of the objects were similar, as shown in Chapter 8. Having reacted to this proposal I was able to show that despite the differences in the angles of tilt, not only were the objects similar, but *all* the main dimensions were proportionally *exactly the same.*

Now, bearing in mind that one set of photos had been taken by Adamski in the area of Palomar Gardens, California, USA, in December 1952 and the Coniston photos were taken in Lancashire in February 1954 by two boys of 13 and 8 years old at the time, the odds against coincidence or conspiracy were totally unacceptable, as anyone who studies the *full* account will discover.

Some people have suggested that such technically corroborative evidence wasn't in the best interests of certain quarters. In other words they may have found it 'inconvenient.' However, as shown in the introduction—*Space, Gravity and the Flying Saucer* was withdrawn in the USA! Strangely, quite a few copies were sold in the UK, some of which found their way abroad, where—according to some reports—the book was favourably reviewed.

Thus the third event was apparently initiated, for I was gratefully surprised to receive from an unspecified source a compilation of notes simply titled "Electro-Vibrationary Forces."[2]

Among other things, the authors made good use of analogies on various aspects of physics which were amenable to repetition with only basic equipment. One in particular, I shall always remember, offered a scaled-up model of molecular motion in solids initiated by the application of heat.

The ingredients for the model are one medium-sized dish, in the bottom of which is placed an ordinary bottle cork. Next, a level quantity of fine sand is poured over the cork to a depth of several inches. Finally, a small lead or steel ball is placed onto the surface of the sand.

In the analogy, one is asked to imagine that the sand grains resemble scaled-up, say, water molecules while the cork and the lead ball represent other molecules of differing densities. In other words, a collection of molecules within a medium composed of other molecules.

If the dish is now vibrated rapidly by some means—at the top end of experimental excellence preferably ultrasonically—the sand takes on the appearance of a fluid. Small wavelets and ripples appear on the rapidly agitated surface. We might view this as if the grains had been frozen water molecules resisting any movement or passage of the cork and lead, but now that the water has been heated—and heat *is* vibration—the denser lead immediately *sinks* to the bottom of the dish and less dense cork *floats* to the surface, where it continues to bob around.

This simple analogy was to illustrate the passage of more or less densely packed molecules through another substrate of molecules by the application of vibration—or heat.

The authors then went on to expand the analogy by showing that electrons can be similarly encouraged to move through otherwise non-conducting insulators simply by the application of vibration, which again is heat. This simple analogy was to set the present author off for several years in the partially successful pursuit of separating hydrogen from water as a clean alternative fuel for use in internal combustion engines. Among the natural derivatives of this work were the submarine ejector engine, shown in my book *The AT Facto,* and other principles, not least wave mechanics.

Naturally, I still have a certain rapport with the alternative fuels quest and, in particular, the chemical decomposition of water to yield hydrogen. Recently, I was reminded of those earlier days by a TV programme titled "It Runs on Water," screened in the UK on Channel 4's "Equinox" series. The programme was appropriately introduced by Arthur C. Clarke in the following terms (which interestingly has much relevance to the motivation for this book). He said:

> I think there is a strong possibility that we are at a turning point in history, a complete revolution in human affairs with the discovery of totally new energy sources. Many people are sceptical of this, but I think we may be going through the four stages involving any revolutionary development.
>
> 1. It's nonsense, don't waste my time.
> 2. Oh, it's interesting but not important.
> 3. I always said it was a good idea.

4. I thought of it first.

Some of the private research work currently going on in the USA was introduced by Paul Czysz, Professor of Aeronautics and former top NASA scientist, who admitted that some rocket engineers were not only interested, but were actively engaged on plans to modify the huge fuel tanks and engines of rockets in readiness to take the water/fuel conversion process. But of no less importance to the point I wish to make is the content of the summary of the programme made by Arthur C. Clarke, in which he correctly added that while such a means of cheap clean fuel for use in engines all over the world would be environmentally desirable, nonetheless there would be an attendant disadvantage with this process in the form of another accompanying pollution—that generated by heat. This is due to the fact that all fuel-burning engines are by definition heat machines, and the more of these there are, the more heat accumulation will build up, thereby adding to the increasing global warming problem. In this context, I might add that currently in the UK the populace is being urgently advised to economise in its use of water—for ordinary purposes, leave alone for powering every internal combustion propelled vehicle—for although an island, the UK authorities haven't even begun to seriously consider the desalination of sea water on a large scale. And even if they do, the process is known to consume prodigious amounts of electricity, and the water produced can cost up to ten times the normal price!

Despite this, if (as I believe) the implementation of the ultimate solution described in this book carries too much of a radical economic upheaval, then for a suitable period of adjustment, hydrogen conversion would logically have to run its natural course. As will be seen, the implementation of the ultimate type of energy we shall be examining will carry no such threat of environmental pollution—heat or otherwise—*provided we are responsible enough to handle it.*

The ever-increasing concern for energy supplies is typified by a statement made by Professor Hiromu Momota of Japan's National Institute of Fusion Science, who stated at a December 1995 meeting that "there is enough fuel in the first 10 feet of the moon's surface to power the Earth for the next 700 years," and that "scientists are hoping to send miners to the moon to excavate the vast store of Helium-3." He said, "Using the moon's Helium-3 will also make it possible to build machines to power spacecraft and even generate high-energy beams to treat cancer." "The moon," he said, "has enough Helium-3 to power the world for centuries; just 25 tons of it could power all of the U.S. for a year." And after that? "Jupiter and Saturn alone have almost infinite reserves."

It is All Around Us

Given the acceptance of the alternative broadly set out in the following pages, it may prove to be unnecessary for such desperate and quite illogical 'moonshine.' For the very source of the energy we require for transport and industry is all around us, beneath the seas, in our living rooms and into the far reaches of space; we can neither ignore nor be separated from it, for the very atoms which form our bodies and brain cells are an integral part of it. Indeed, we are truly of one and the same origin!

At present, none of us can be certain how long it will be before the power predicted here will be generally identified and successfully utilised, but based on research, I can state quite unequivocally that it *does* exist and it *can* be utilised. Moreover, it is the prime intention of this book to offer evidential directives that this is truly so.

My investigation has not been without results which on occasion produced amusing and dramatic effects! But for some obvious reasons and some not so—dealt with later on—we can only discuss the concepts involved fundamentally here. Even so, the discerning reader will find persuasive directives sprinkled throughout the text.

A Beginning and a New Approach to an Old Problem

I have considered that the decision to publish this work is not without merit, if for no other reason than fairness to my wife, family, friends and, not least, all those who are feeling depressed by the otherwise uncertainties of the world's environmental future and energy reserves. Though I have to concede to the view that given its availability, it is questionable if the principle herein outlined could be realistically regarded for our use *in the present era.*

However, nowadays it is practically impossible to go one week without being reminded by the media of the world's pending energy and ecological problems. Twenty years ago the spelt-out messages were there just the same almost word for word, the coming non-renewable energy crisis, acid rain, the greenhouse effect, etc.; what is more, the predictions are coming true and they are likely to go on doing so as we approach what some think will be the beginning of the deadline—the year 2000.

Millions of pounds and many more man-hours have been, and will be, spent on alternative energy studies in an effort to help alleviate the day when children en masse could otherwise go hungry in the industrial world as well as the underdeveloped countries. Such unpopular information should not have to be reported in a book such as this, for there are plenty of far more adequate reference sources elsewhere, but within the context of the prediction made in these pages, it is vitally important that we should constantly be mindful of the possibility as we proceed.

Most engineers pondering on some of the pending crisis forecasts find themselves wondering why we can't do this or that, it's a natural challenge which often brings pen to paper and many bright doodled ideas. In fact, there are scores of patents filed every year concerning renewable energy sources—involving wave, tidal, wind and solar energy, etc. Unfortunately, although these ideas are based on well-established principles, if put into practice they will only help to *ease* the problem, *not solve it*, for most of the predictive calculations are based on the world's energy consumption rates of the *present*, not the year 2000-plus. As will be seen in Chapter 2, at the present level of development, the *rate* of consumption will be considerably more in the year 2000 and the decade *after that*. Having a keen interest in the enigmatic nature of gravity, perhaps it was natural that I should have become an early member of the British Interplanetary Society, at which time we had troubles other than environmental ones to contend with . . . World War II. At that time, of course, I had no idea that such an aerospace-directed pursuit could inevitably lead to identifying an alternative inexhaustible energy source. Being an aerospace engineer rather than an astrophysicist, I was able to relate more comfortably to classical physics as laid down by Newton, involving visual concepts rather than relativistic or quantum ones. Not surprisingly, therefore, what productive efforts I may have contributed to the nature of space and gravity were in traditional Newtonian terms. Also, to some extent it may explain why the tenor of this book has been largely coloured with an aerospace flavour.

However, such a would-be discoverer in this somewhat alien land is as much ill-equipped as a diver without an aqualung and you very soon have to come up for air! But there is nothing to prevent you from diving in again and again for many short dips as I did. Nevertheless in no better way can the orthodox summation of such furtive excursions be illustrated than by quoting from the family copy of *Pears Cyclopaedia* which offers this conclusion:

> Over a century's development of the atomic ideas has brought a progressive, if jerky, increase in the mathematical precision of the theories. In some fields of particle physics, observations to one part in a million, or

even better, can be explained, to that level of accuracy, by the existing theories. At the same time however, the theories have lost visual definition. An atom as an invisible but none the less solid billiard ball was easy enough; so was a light wave conceived like a sound wave in air. Even after Rutherford, an atom consisting of a miniature solar system merely exchanged the solid billiard ball for a system of revolving billiard balls and was no great obstacle to visualisation. But since quantum theory and the Uncertainty Principle, every unambiguous visualisation of fundamental wave-particles leaves out half the picture, and although the electrons are in the atom, we can no longer represent them in definite orbits. The moral seems to be that *visualisation is unnecessary, or at best a partial aid to thought.*

Now if like the author you *have* succeeded in developing a feasible visual conception, the obvious logic in the above can be very daunting, for it is apt to leave one in a kind of in-between no-man's land, where you cannot shake off the persuasive validity of your visual model on the one hand, neither shun the alternative—and to a large extent—established views on the other. Thus it was for many years that I accepted this impasse and got on with my other research in aeronautical ventures, quite convinced of the fact that while eventual developments in astrophysics *might* lead to a solution to gravitational manipulation, as to produce zero gravity and beyond—of the kind in fact which my visual model anticipated—I had been equally convinced that in the meantime rocket motors were the only alternative for space flight.

I had yet to discover, however, such exertions would indeed point the way to a possible exciting alternative space drive. That such a principle would inherently embody so-called 'free energy' could hardly have crossed my mind, for such an extravagant proposition would require one to believe that an aerospace engineer, working solely from a somewhat fundamental visual conception, might evolve a very simple theory pointing the way to a combined space drive/energy production system which—some have said—could have dramatic effects of 'quantum leap' proportions, whereas physicists had not. Therefore, if at first hearing I could not have believed this possible, I must now expect others to have the same reservations. However, it is hoped the material in this book will prove more acceptable when considered in the light of such observations as the following made by an eminent scientist of the day.

In an article by Philip J. Hilts of the Washington Post (1991), Leon Lederman, director of the Fermi accelerator (the most powerful atom smasher in the United States) was quoted as saying, "We have reached a crisis in physics where we are drowning in theoretical possibilities *not based on a single solitary fact.*" The article went on to say, "New questions are being asked about what matter is, how it is formed and why it behaves as it does. Physicists do not understand why there are so many types of fundamental particles or why they vary so greatly."

Now if such technological developments as those described later would be enormously exciting, then some of the associated implications could be significantly negating. To achieve such a goal is one thing, for it to be available to a world culture hardly capable of sensibly looking after its domestic affairs is quite another! Moreover, it would be naive to hope that man's attitudes would improve in step with such technological advance, yet the alternative may be too dreadful to contemplate. Therefore, the ideas expressed here are offered in the hope that investigation into this technology will continue to fruition—as I have reason to suspect it might be elsewhere— and be available when necessary to a more adaptable and enlightened world. Meanwhile, in order to achieve that I think we should be well advised to regard our present position as tantamount to a third world war situation, not as is generally accepted between the so-called super powers, but with the enemy fragmentally dispersed among society which depends on people's ignorance, fear and frailty to win.

In a word, while under the pressure of significant world events, humanity can respond flexibly to new ideas, the moment the pressure is off it all too often resists energetically the need for fundamental change. Therefore, the aerospace industry, for instance, is hardly likely to be seriously investigating the type of propulsion the author predicts here.

Emphatically, this book should not be regarded as being merely a science fiction-based fantasy, although it is accepted there may be those who would prefer to interpret it as such. Rather, this is an attempt to reveal that much of what people regard as science fiction of tomorrow, may incredibly already be established science fact and this work should be regarded as a serious message to anyone who has the capacity to think and assess this for themselves; the only degree required of the reader is belief and patience.

It is also anticipated that there will be scepticism among some elements of the scientific community, but we may take courage from the knowledge that on an occasion known to the author, when a well-meaning enthusiastic inventor staked a claim for economically reducing water to fuel (chemically) he was —outwardly—not taken seriously. Nevertheless, unofficially, members of an august fraternity were surreptitiously quick to attend a meeting for a 'know-how' discussion. I have often found this attitude difficult to reconcile, unless of course it is a matter of *just in case!*[3]

Neither is it only the amateurs who may let their imaginations get the better of them. Based on the finding of a poisonous 'fang', it was reported that scientists recently worked out a whole lifestyle for an unknown 100-foot-long giant serpent in South America. Casts of the fang were taken and sent to museums and universities around the world before someone identified it as a piece of ordinary seashell!

Fortunately, the majority of scientists from all disciplines usually demonstrate more restraint. In fact, lay people are apt to be unaware that most scientists are so heavily committed to specific tasks in industry involving *applied* research, that they have little time to devote themselves to *original* research of a revolutionary rather than evolutionary nature, which is therefore usually left to a few. It is accepted that this has to be the way of things.

However, in addition to some doubting physicists and engineers, it is anticipated that this present work will produce more than a little disquiet among some ministry officials, industrialists and militarists to mention but a few. That being so, I am encouraged by the further anticipation that my efforts may be received more favourable by more socially oriented bodies such as Greenpeace, Friends of the Earth, and the Noise Abatement Society, etc.

It is becoming obvious that new methods of transportation/energy will probably *have* to be adopted as have ships, cars, trains, aircraft and spacecraft in the past. We must now accept any meaningful progress or face the grim prospect of our species' ultimate extinction. It is known beyond a shadow of doubt that with our present technology mankind's future stay on this planet is highly questionable indeed. I repeat, let those who remain in doubt check the status quo for themselves. Bluntly, the message that ecologists the world over have been trying to spell out to the rest of us can be translated thus: In its final analysis, the energy *and* pollution crisis should be regarded with no less awe than does the prospect of an all-out nuclear conflagration; to which we should perhaps add, it is not so much the—by comparison—almost childish issues of 'nuclear missile balance' and 'Star Wars' defence systems about which the leaders of the world should have been arguing, but given present technological standards, how long have we left for us to have a world and population to defend?

The material in this chapter has been deemed informative and necessary in order to acquaint the reader with the fact that there really is inherent in every society certain obstacles and elements which automatically act to curtail important techno-logical changes—more succinctly described as 'inertia'—and these have to be vigor-ously overcome. It has been considered the best way to achieve that is by this kind of exposure. Also, it must be reemphasised that in view of the controversial nature of some of the predictions made in the context of a vast natural energy field, it may not only be useful but necessary to include along with the specific format some of my supportive background experiences as well, if for no other reason than to offer the reader some credibility and even introduce a little humour into an otherwise serious subject. Certainly, in no way should this material be interpreted as banner waving; the issues at stake are far too important for that. For instance, as it will be seen later, I have spent a good deal of time working on jet propulsion gas turbine engines and have produced several ideas, one of sufficient merit as to be taken up by one of the world's largest aero-engine manufacturers. Therefore, I feel qualified to state un-equivocally that despite their magnificence, modern aero-jet engines are among the most complex, most expensive, and (more to the point of this book) in aviation roles, the greatest fuel gobblers known to man. However convenient or however much ben-efit mass air transport has given to us all, it is a sobering thought to consider that every time a jumbo jet crosses the Atlantic, in terms of fossil fuels it burns an equivalent of several acres of forest trees! Yet, despite this, we can expect and understand resis-tance to new ideas. For example, we would not be hard-pressed to imagine the reac-tions of, say, Pratt and Witney, Rolls Royce and the establishment generally, to the realisation of the principle discussed later on, which requires but little starting energy and few moving parts. Certainly we would hardly expect them to clap their hands for joy! Predictably, therefore, any discussion concerning such work will be received with a mixture of incredulity, alarm and the fervent hope that it *is* just a joke, which although understandable—bearing in mind the seriously limited extent of the world's renewable energy reserves—is rather sobering. Even so, whether the principle herein presented will ever be allowed to be adopted or not, I suggest that this book might be received as a fair declaration of the reality of what I have termed Comat energy and mankind's capability of harnessing it for all our needs.

As an aerospace engineer, my work has often taken me into areas of establishment sensitivity and, perhaps not surprisingly, at lectures I have been asked why the natural development of such aircraft as the Concorde has been concluded and why the Americans dropped out of the scene?

During my time, I have come across speculative answers to that question on many occasions; perhaps a more recent quotation from one of aviation's best-known pioneers is appropriate. In a November 1994 TV programme screened by the BBC, Channel 2, titled "Perpetual Motion," Brian Trubshaw (Concorde's first test pilot), concerning the closing down programme of his much beloved aeroplane, said:

> Transport has always sought to decrease journey times right from the days of stagecoaches and clipper sailing boats, steamships and aeroplanes and for the last thirty years aeroplanes haven't gone any faster except for Concorde. Now one of these days men of vision will do a successor to the aeroplane; sadly I don't think the initiative will come from this country, it will be a big international project involving America, Japan and possibly France as well.

Facts like these have weighed heavily in the decision to publish my findings somewhat prematurely. Indeed, over the years, others have tried to persuade me that among this background material there would be sufficient information of general interest to fill a separate book. Indeed, at one time something on those lines was professionally bandied about a little, a tentative title being *So You Want to be an Inventor* or something like that, which would have highlighted some of the pitfalls, excitement and disillusionment therein.

To have done so would have meant releasing the enclosed work on Comat energy separately and to a much restricted readership, which would have considerably negated my prime intention. Whereas presented as it now is in a condensed, supportive and hopefully more generally interesting role, a greater awareness of the importance of some associated issues may have a much broader appeal.

For instance, it may be a sobering thought to consider, that if the author's claims are correct, then the tiniest fraction of the mind-boggling collective sums now being spent on a wide spectrum of diversified endeavours, ranging from ever-increasing miles of motorways, more and more expensive fuel-guzzling motorcars, air and space craft, research into *known* alternative energy sources and remedies to minimise worldwide environmental pollution, etc., etc., were to be spent on research into identifying the colossal energy source forecast, it might just yield *one simple all-embracing solu-*

tion of the kind presented in this book.

"Wood for the Trees" as a meaningful subtitle to this chapter was chosen after due deliberation for the need to set the stage for what after all might prove to be a controversial issue. It may also be interpreted as an attempt to express gratitude to an eminent aerospace scientist who had the courtesy to be absolutely honest when confronted by a comparative novice with—what might have been—a screwy idea. In an ideal world it would be reassuring for this to be the norm, but regretfully many of us know from experience this is not always true. However, the forthrightness of the one can often compensate for the transparent bigotry of many and it is hoped the inclusion of the following introductory account may help to establish that stance throughout the remainder of the book.

An Aeromodeller's Introduction to Turbojets and Surface Effect Vehicles (Hovercraft)

In 1942 I was directed by the Ministry of Employment to the De Havilland Aircraft Co., at Hatfield, Hertfordshire, as a technical illustrator; yes, they were glad to take on almost anybody in those World War II days! I had spent some time in the Army attached to a Royal Engineers Unit digging up Hitler's delayed action bombs . . . (I have never been quite the same since!). This was followed by a post with Taylor Woodrow as a trainee junior civil engineer, due solely to the fact that I could draw fairly well and had an eye for the dumpy level and theodolite. Most of the time I spent helping to set out the runways and perimeter tracks on new aerodromes in Norfolk; chiefly Snetterton Heath. During this time, and after they were completed, I spent as much time as I could indulging in my love of aircraft, sometimes just looking at them and studying wherever I could. So in hindsight it seems strange that later the Ministry of Employment should quite coincidentally have pointed me towards Hatfield as they did. It would prove to be the beginning of an aerospace adventure for the rest of my life.

Aircraft Gas Turbines and Panic Alert at Hatfield

I immediately recognised the first turbojet engine as something I had imagined as a youngster, for somehow I had never really felt comfortable with the idea of a 'stick with a large fan at the front,' though I appreciated the need for it. Constantly being away from home no doubt helped to fire a fascination for gas turbines which occupied a good deal of my spare time and gave wings to an intense wish to design and build a miniature version for a model aircraft. However, this was extremely difficult to achieve as I was living in lodgings at the time, but the chief experimental engineer at De Havilland's—no doubt recognising my enthusiasm—allowed me to use a friend's workbench in the corner of the vast hangar which housed the first, then top secret, Vampire jet. I well remember the day when in the lunch hour with the help of several engineer friends, we ran my small turbine of some three inches diameter, up to operating speed of about 60,000 rpm (revolutions per minute) to check it for balance.

For a given power the smaller the diameter of a turbine the higher must be its rpm. Therefore there can be a similar frequency of the number of blades meeting the gas flow per second in a small turbine as there may be in its full-size counterpart. In a word the resulting pitch of the turbine scream can also be very similar. In this instance of course we were running the unit up with a 60 lb. per inch^2 pressure air line.

Now in the comparative quiet of a huge lunchtime hangar . . . the noise emitted from this baby turbine was horrendous! So much so that it produced an inrush of startled security police and personnel quite convinced that such an unscheduled run-up of the Vampire 'goblin' jet engine implied some kind of trouble, if not virtual break-in! (This small engine which I designed and built in 1942 is shown in the Appendix.)

First Hovercraft Demo

But of no less interest, it was primarily due to working on this little jet that I first became acquainted with the surface effect of so-called hovercraft. This is the way it happened.

A small gas turbine engine can be very inefficient, in fact it is almost impossible for it to run self-sustaining. However, it is also known that a pulsing gas flow jet rather than a continuous one can help this lack of efficiency. Indeed, it is the very same principle as that employed in the German V1 'doodle bomb' jet.

I had figured that there was some merit in employing the principle for the small gas turbine, but some experiments were called for. One such arrangement—shown in the Appendix—was investigated in 1943 in my landlady's kitchen. This was fed with gas from the good lady's cooker (while she was away of course) and was specifically aimed at checking flow rate and the ignition system via a flap valve. The theory being, the weight of the unanchored dish would allow it to pop up and down with each resulting controlled explosion. Subsequent guesstimation placed the frequency of this to be something in the order of 50 cycles per sec. But for one very brief predictable moment the dish floated and skimmed around the table! Fascinated, I had contemplated this phenomenon for a little while before resuming "the more serious work," completely oblivious of the fact that I had just been introduced to my first plenum chamber-type hovercraft, which years later would involve so much of my life.

Sequel Panic on the Isle of Wight

What happened to the little turbojet? Well it *was* completed but never ran continuously and still resides in my workshop. Moreover, as a gesture, I was presented with a complete general arrangement (G.A.) of the then-secret Vampire aircraft from which I built a free flight model to take the jet. This model still flies to this day; moreover it is so realistic that on one occasion, nearly four decades after the occasion at Hatfield, with smoke pouring from the tailpipe it was seen to make an 'emergency crash landing' by several dozen cliff top holiday-makers who were totally convinced that they had just witnessed a full-size aerial drama! I, for one, have become accustomed to this strange phenomenon of 'synchronicity.' By the way, through what persuasive channels did I manage to acquire the Vampire G.A.? It was none other than the young David Kossof, who would one day become a world-famous actor!

"From Little Acorns Grow"

In 1948 I joined the design office at Messrs. D. Napier & Son Ltd., at Acton in London, U.K. As is well known, their speciality was engines, ranging from the large 'Deltic' compound diesel engine—used all over the world in diesel electric locomotives—to smaller, more powerful turboprop and jet engines.

1956 found me attached to the advanced projects department (APD) where we often had to graphically complete certain areas of the designs which required rapid attention in order to meet a deadline date—usually tomorrow—for important meetings with various dignitaries from such as the MOD (Ministry of Defence). One design designated the HTU1 (Hot Turbine Unit) involved a rather costly, convoluted and complicated solution for cooling turbine blades by a process of forming them hollow in order to allow cooling air to be pumped through them, a process which involved conveying the air through a very clever series of labyrinth seals. The idea was sound enough but extremely tedious to graphically portray—we were not blessed with graphic displays and computers in those days.

In view of the short time available, I was not overpleased with this somewhat daunting task, nevertheless it did have the effect of spelling out to me the extreme lengths designers were prepared to go to in order to increase gas turbine working fluid temperatures (WFT) by only one or two degrees Celsius.

On my journey home I was pondering the problem in terms of some of those I had met years ago. I found myself contemplating some of the technical compromises gas turbine engineers had to make and if the turbine power *output* could be *considerably* increased by only raising the WFT by a degree or so, just imaging what output could be achieved from the same engine if the WFT could be significantly raised! Indeed, suppose the flame temperature need not be diluted *at all*? My mind boggled as I thought "if only they could do *that!*"

Of course, by this time I was sufficiently skilled in the art to be able to take the thought further with a few basic sums and some rough sketches. The conclusions were disturbing, for I couldn't fault the idea. But I was equally impatient because it was so obviously simple, engineers would have tried it all over the world, for if my hypothesis was correct, it implied jet propulsion turbines could in fact be formed from light inexpensive alloys, for with such a design they could *run cold*! Sometime I would curb my pride and ignorance and try to find out. That time turned out to be sooner than I thought.

The following day I had to direct all my attention to getting the job on the HTU1 done in order to meet the deadline. But this was not easy, for working on the complex system only served to rekindle the thought: "why . . . why not?"

During the morning Mac, the technical writer working on the job with me, came over to discuss matters several times and I found myself impatiently wishing to put the question to him, for he had a doctorate in thermodynamics and was certain to be able to quench my disquiet. As the morning wore on, I finally yielded and apologising for the digression, showed him my rough notes with the question, "as there *must* be a reason, what was it?" This was followed by a somewhat disturbing silence and a brief moment or two of hesitance, then finally he said, "I don't know!"

Several days later we had finished our allotted job on the HTU1 in time, our visitors had returned to the city highly pleased, our boss was relieved, and so were we.

Not One but Three Thermodynamicists

Around three o'clock in the afternoon, Mac came in to see me rather excited and slightly flushed, I thought, and said "With regard to your question the other day, per chance this lunchtime I sat down at a table occupied by not one, but three APD thermodynamicists." He paused, then said, "I hope you don't mind, but I put your question to them." Sure I didn't mind, if anything I was glad to have a chap of his technical calibre at least willing to be told, as much as myself. I waited patiently, then in response to my questioning glance he said, "That's just the point, it's now three o'clock in the afternoon and I have only just left them . . . *arguing among themselves* about it!"

For several nights I worked long into the small hours, as I have always done, building a fully sectioned table top display model engine complete with rotatable rotor and pretty stand. I had a hunch I was going to hear more . . . and I did.

Breakthrough and Dr. Morley

Within a day I was called to the 'higher sanctum' in the firm's technical division at the request of the then-chief advanced projects scientist, Dr. Morley, on assignment from the Farnborough Aerospace Establishment. My own immediate superior was enjoying himself, bathing a little in the 'reflected glory' as he gave me the news: "Just imagine one of his chaps with a major breakthrough potential."

Dr. Morley's secretary announced my arrival and I was cordially greeted by him sitting behind an impressive desk in an equally impressive room with walls hung with a variety of pictures appropriate to the aerospace industry. He rose to meet me, shook hands, and proffering a chair, came around and sat on the edge of his desk lighting a pipe in the process. Then, after formally enquiring about my background the questions began: How long had I worked in the projects department and what motivated the idea?, etc. Dr. Morley seemed curious to know as much about the cold turbine's inventor as the concept itself.

Eventually, I handed the box containing the display model to him and one of the fondest memories I cherish is the broad grin of appreciation on his face when he raised the lid, and I was immediately glad for the long night hours I had spent building that model. He duly thanked me for it, saying he would be delighted to use it at future meetings as he had already placed the idea in the hands of several jet propulsion specialists in the APD. However, I was entirely taken aback by Dr. Morley's next question when he asked if I would be willing to join his team in the APD as he considered I was something of an 'ideas man.' I cordially thanked him, but had to tell him my family and I were considering moving to the Isle of Wight, and in any case, as higher mathematics was not my forte I might prove to be inadequate for such a position. He said that my intended move was a pity, but as far as my rusty math was concerned he would be glad to make arrangements for me to polish up at Napier's technical college a couple of times a week.

"The Wood for the Trees"

During the period which had elapsed since my first conversation with Mac, I had little time to devote to curiosity, yet I was puzzled somewhat insofar as surely my proposal was more the province of specifically trained engineers? Leaving Dr. Morley's office that day, something he said made me hesitate with that unanswered thought still lingering in my mind. I shall never forget the image of that kindly man as sitting there with arms folded and pipe in hand. He, too, hesitated for a moment, then smilingly said, "It's simply a matter of all too often we can't see the wood for the trees!"

A new future had been offered me, a self-taught aerospace enthusiast, and although it was a tough decision to take, my family and I did move to the Isle of Wight where the pull of a job working on Britain's first space rocket was too much to resist. I would have been honoured to work for Dr. Morley and share with him some of the magical things it has since been my good fortune to find.

It is my sincere wish that this short account may in some measure help to reassure other equally not so well trained, but insight-gifted youngsters not to be afraid to ask "Why?", as this retort from an honest man may convey. Since those days, I have seen the truth in that stance upheld many times. However, from time to time we all suffer from "wood for the trees" myopia. I certainly know I have, and the truth in that metaphor has not weighed lightly in the decision to present results of my work on gravity and energy in the form of this book.

Liquidation and Missing Documents

What happened about the 'cold turbine' jet engine? The Ministry of Defence immediately clapped a top secret notice on it. I met with officials several times. Then the project was taken over by a subsidiary firm, English Electric at Luton Airport, Bedfordshire, U.K., for wind tunnel tests where it was eventually shelved, *as the faculty had no access to the required hypersonic wind tunnel,* which was then unavailable in the U.K. Eventually D. Napier & Son Ltd., went into liquidation, therefore in hindsight I did the right thing in resigning my position and moving to the Isle of Wight. Subsequent events deemed it to be so.

I never heard another word about my cold turbine design, although Mac did manage to rescue a copy of the patent application—shown in the Appendix—together with the design study from the APD. Some years later this was found to be missing from my workshop file, following a visit from representatives of a large overseas concern who were showing interest at that time in one of my other projects!

Science and Prejudice

Having received my share of prejudicial reviews concerning some of my earlier work, I naturally anticipate repetitious debunking from certain quarters now. It is not unusual, however, for people to eventually change their minds as new facts emerge. Indeed, among my best friends are several physicists from the scientific community who have done just that. But intransigent prejudice in the face of available evidence is quite another thing, as I know to my cost.

In 1966 my second book, *Piece for a Jigsaw*, was published. Essentially, it was an extension and update on my first book, which as stated consisted of an unbiased report on some extraordinary similarities between many UFO incidents and my theory of gravity wave-propelled spacecraft (repeated in this present book in an abridged form in Chapter 5). As indicated in the Introduction, it suggests that given the required technology of this form of 'propulsion,' then purely on sound engineering grounds we can quite accurately predict some of the associated physical effects we might expect, i.e., surface and atmospheric, etc., and I filled the entire book with these and still left many more to the reader's imagination; one obvious example will suffice. The theory suggests that if a machine of this type takes off from, say, a field, then the pilot must do so slowly, for to increase power and ascend too swiftly implies that despite the inverse square law, the 'field of force' would probably extend beyond the boundary of the vehicle, most likely causing some ground in the vicinity beneath to 'fall' upwards with it, thus a crater would be formed. This can quite easily be calculated and I showed this in the text.[4] I told how I had noticed such identical characteristics among UFO incidents which had been described by ordinary people, ranging from housewives and children to doctors and scientists.

The proposition I put forward at that time was reasonable and simple to understand, i.e., the statistical odds against this and many other aerodynamic effects being coincidence were totally unacceptable. Therefore, on purely engineering grounds, I had to accept that when analysed *in terms of a simple theory*, many UFO reports made perfectly good scientific sense. It was interesting to note that many of the normally highly vocal dissenters became very silent, while today there is an ever-increasing number of scientists who give me their support.

However, in this respect, I can philosophically reflect I have been in good company, for the records show that in the 1930s Sir Frank Whittle and his (at that time unknown) German competitor were treated identically by the then-entrenched aeronautical engine manufacturers who continued to totally refute the viability of the jet engine concept, which has since contributed to mass transportation of people who would otherwise have never visited other parts of the world! Here are a few other timeworn instances of professional bigotry:

1878. Professor Erasmus Wilson on the first demonstration of electricity said, "With regard to the electric light, much has been said for and against it, but I think I may say, *without fear of contradiction*, that when the Paris Exhibition closes, electric light will close with it, and very little more will be heard of it."

1904. Lord Kelvin, president of the Royal Society, said, *"Radio has no future . . . "*

1958. Thomas Watson, IBM chief executive, said, "I think there is a good market *for about five computers.*"

1957. Sir Harold Spencer Jones, Astronomer Royal, said, "I can say quite definitely and with absolute assurance, that none of the flying saucers can have come from another planet!" But he is equally well known for stating that *"Man will never set foot on the moon."* And wasn't it Sir Bernard Lovell who said that *"All space flight is utter bilge"*?

It was yet another all-knowing soul who stated on Albert Einstein's Munich High School report, *"He will never amount to anything. "*

That is prejudice, and the whole of science and technological development is plagued by it.

Why the Paranormal?

I shall be presenting facts as I have found them, and as a British aerospace engineer with over forty years experience in forward research projects, together with quite a few patents behind me, I feel qualified to do so. However, due to the nature of some of my experiences, I quite accept that there will be those who, being unacquainted with some truths (about so-called paranormal phenomena, for instance), will be apt to prematurely 'label' the author and will not take the subject of this book seriously; that is regrettable, for they will be missing a vital contribution. For my part, I shall always be grateful for such early enquiry, without which I doubt if these facts would now be recorded. Therefore, although in some circumstances it would be unnecessary—and in some others not even wise—to elaborate further with details of personal experiences, the nature of the prime directive of this book suggests it would be remiss of me not to do so, as will become clear later on.

As a youngster—no doubt along with many others—I had an intuitive acceptance of so-called paranormal phenomena, but it was not until the age of twenty-five that in the reliable circumstances of a private home and in the company of my family and trusted friends, that I witnessed genuine levitation. I was to learn much later that to some paranormal researchers this was almost the norm. To them this was exciting? Yes. Bordering on the miraculous? No. And with respect to lay persons, that was all they felt about the phenomena, just another levitation!

Now *any* aircraft engineer would be *very* impressed, considering the energy and fuss normally required to lift a 30 or 40 pound weight, which under specific circumstances can be made to float silently like thistledown, for the expenditure of . . . what? I reasoned, *if* it is *energy* as we know it, then it is so small as to defy all attempts to measure it by conventional means. But although at that time I was a lay person in such matters, I was also an enthusiastic student engineer with 'new' jet propulsion gas turbines, space flight and rocket motors very much in mind. To say the least, I was rather alarmed and not a little impressed, while resolved—for the moment—to keep this very much to myself, chiefly where aircraft engineers were concerned. The following may serve to illustrate why.

Difficult as it may be for some younger aerospace engineers of this era to believe, to this day I can look back with some disquiet, in memory of the sidelong glances and almost alienating feeling measured out by our colleagues in the design team at the De Havilland Aircraft Company in Hertfordshire, UK—when they discovered that together with a mere few, I was a member of the embryonic British Interplanetary Society. *Truly*, at that time we were considered rather outlandishly *eccentric* and were churlishly patronised. But before they are inclined to raise eyebrows with incredulity, may I remind some present members of that branch of the establishment that *they too* may be just as capable of biased disbelief in something which may be beyond their present comprehension. If they doubt the validity of this, now is the time to find out!

I had joined the company towards the end of World War II, after leaving Taylor Woodrow to work on one of the first jet aircraft in the world to go into service with the RAF. It was there that I was first introduced to Hitler's 'doodle' bomb or V1 GCM (gyroscopically controlled missile). But it was not until the first V2 bombs landed on London and we heard the explosion *followed* by the slip stream noise from the missile's free fall descent passage, did I and my BIS friends realise that we had a supersonic missile to contend with. Perhaps, predictably enough, those sidelong glances in the drawing office ceased, but any sense of vindication on our part was nullified by the impact of what we knew to be true: in another sense we would have preferred to be wrong!

I beg the reader's indulgence to be allowed this little reminiscence, but its relevance is appropriate. For it so happens that in the ensuing years, I have experienced those same sidelong glances many times, and I pen these words in the full knowledge that they will now come again, only this time in greater numbers. But one thing is certain, I can assure the reader. If you are an aerospace engineer worth your salt and you observe *genuine* levitation phenomena, you are not the same person the following day! To the reader who on this account alone may now be tempted to close this book impatiently, I say confidently that not only do so-called paranormal phenomena occur, but that their *origin* will be the means which you or your future family may one day employ for transportation and hopefully total energy requirements.

Bearing all this in mind, it is perhaps predictable that if you happen to be a keen space flight enthusiast, such experiences are most likely to influence your thinking. It certainly helped to motivate the author in trying to understand the nature of gravity.

However, as an aircraft design engineer more than a physicist, I was relieved and encouraged to find that if I knew nothing about this celestial phenomenon, then self-evidently—apart from describing and predicting certain gravitational laws and *effects*—nobody was offering me a satisfactory visual model either.

I could quite easily fill another book with accounts of my earlier theoretical and experimental meanderings over the intervening years, involving 'space dynamic gradients' and 'push' from without rather than a 'pull' from within theories ad nauseum, and sandwiched as it had to be between my occupational work in aerospace, this has incurred considerable expense and time. At lectures on the subject of gravity, I often used the analogy "the only energy required to sail a boat was consumed by the sailor in setting the sails, thereafter the energy being supplied by the wind, until a change of course was required." I often voiced the hunch that one day men might travel on 'currents' of space like that, perpetually, the only energy required would be tantamount to setting the sails in the analogy and that such a dream was not of this age, belonging to some far-off future. I now know that only in the latter was I wrong.

The Law, Patents and Need for Moral Change

Our eventual successful adoption of the main topic presented in this book will largely rely on our humanity towards one another, i.e., we would *have* to be a responsible society. In a word there are many interdependent factors which will have to be implemented along with any dream of a new world technology. Swift changes will have to made in law in some vital associated areas, whereas at present there are all too many glib excuses made for blatant legal 'inertia' in some more obvious areas in need of change.

Despite our fond belief that we live in a just and benign society, many are aware that this is far from the truth. The truth is, that we live in a society which is just and benign only when compared with somewhere else! For instance, stripped of the utopian promise of unlimited energy, space exploration and plenty for all, we unfortunately have to remind ourselves that we still live in a society which tolerates—among other things—appalling unfair legal situations involving industrially maimed and injured people, who after many years, having won their case, too often lose an extortionate proportion of this award in 'legal' costs. One would like to think that couldn't happen in *our* country, but it does, and in the world of gifted original thinkers, innovators and inventors, the indifference and unfairness is exactly the same.

Many people are under the impression that any inventor can apply for patent rights on almost any idea and obtain 'protection'; unfortunately, this is a complete fallacy. Although it may be denied, there does exist a patent selection board whose prime directive is to identify ideas which may be 'sensitive to national security.' That being so, it's not a bad idea when you really think about it, but knowledge of it might prove very daunting to private researchers with ideas about energy and propulsion with significant implications.

Also, as we are all now aware, it is not without precedent for *literature* containing material 'sensitive to national security' to be suppressed or withdrawn and it might be pertinent to suggest that this doesn't carry a special exclusion clause for inventors.

Indeed, very few people are aware of the fact that should they—in a private capacity—come up with a revolutionary idea which officialdom considers might be used for military purposes (defence of course), it can actually be confiscated. The inventor would have to relinquish all further rights to it forthwith, despite the fact that they could have spent every penny they had and a lifetime developing it. In some instances the official attitude seems to have been, "we don't understand this thing and we don't want it anyway, but neither can you have it back, just *in case* it could be useful to an enemy." I have knowledge of several such cases, one involving a very well-known inventor who was responsible for a transport vehicle now used world-wide, which was 'shelved' *for four years!* Such is the type of restrictive thinking that some—to whom the taxpayers of countries like ours pay very high salaries—are capable.

Moreover it seems that when it suits them, they have been 'ordained' with even more extraordinary powers of 'scientific' authority, for an inventor from Lucedale, Mississippi, Mr. Joseph Newman, has been engaged in a seven-year controversy as to his right to be granted a patent on his claimed energy producing machine. It seems the U.S. Patent Trademark Office (PTO) has vigorously refused him a patent on the grounds that "it smacks of a perpetual motion machine" (which by *present* scientific definition isn't possible).[5] Later on I propose to show that this statement should be extended to include "according to the limitation set by classical Newtonian physics," and further, that not only *is* so-called perpetual motion possible but—within the framework extended by the findings presented in this book—*it is absolutely predictable.*

Unfortunately, I am sure that nowadays it is generally conceded that without substantial funds and/or backing, it is practically impossible for most inventors to realistically protect their invention from large industrial concerns. I make this assertion not without considerable experience, not least, the disquieting memory of the occasion when my son and I were offered a fair-sized contract by a world-renowned company merely to 'get round' an equally well-known competitor's existing patent! And on another occasion, when a product that had taken us nearly a year to complete, was conveniently 'lost in transit' due to the fact that the large company involved was—unknown to us—already deeply committed to a similar venture!

In this respect one also has to come to terms with the fact that once a comparatively simple process (for hydrogen separation for instance) has been brought to patentable fruition, it will be extremely difficult to meaningfully protect, for its duplication will be practically available to every competent car-driving engineer. Under-the-counter kits will immediately become available; in much the same manner as private builders started to build their own small hovercraft without patent levy to Sir C. Cockerell years ago. And in this context, as will be shown later, I am totally aware of the fact that if it is eventually verified, the main topic of this book will prove to be no exception to these restrictions. Moreover, it goes without saying that if water as a fuel—or any other advanced propulsion/energy process—became available, there could be very stringent military interests imposed.

Intrigue in the Establishment

Apart from all this, it should be stressed that it is not only official 'inertia' with which the innovator is frequently confronted, but equally juvenile establishment behaviour also—all too often by people in fairly high places. A more notable case brought to my attention by a leading scientist was his claim that his immediate superior openly *cooked* his reports and *borrowed* information from his team in order to further his efforts toward a knighthood and fame, which he ultimately achieved and became very well known as an influential and respected authority! Establishment 'law' and prejudice is an internationally established fact; not for nothing was the adage N.I.H. (not invented here) syndrome coined by suffering candidates.

On an occasion when discussing some of my own earlier independent work on Surface Effect Vehicles (Hovercraft) with Sir Christopher Cockerell, he recommended that I should be advised to offer the work to the large company involved with this concept, with the caution that "don't be surprised, you'll probably get a raspberry just as I did." The persons to whom Cockerell was referring were later to become internationally respected as foremost hovercraft designers of one of the world's largest hovercraft companies! This is so much water under the bridge now, and I am sure Sir Christopher will allow me indulgence by quoting him in this respect. In truth, the matter only reflects a modicum of the frustration suffered by this brilliant man. By the way he was absolutely right . . . *I did!*

The Visiting Generals

Perhaps in no more emphatic way can I convey to the reader some insight into intrigue of the 'inner sanctum' than by indulging myself one other quite true morsel. It has much relevance to the main subject of this book. In 1956 I had newly joined the above company, working on space vehicles and advanced fighter aircraft projects. Prior to joining this concern, however, I had developed my field propulsion ideas to a sufficient degree, which I was able to publish in my book *Space, Gravity and the Flying Saucer*, and several magazines. Having settled in at the new company for several months I was already aware of the fact that great overseas interest was being directed towards our new mixed propulsion unit fighter projects; *visiting VIPs from other countries was almost the norm!* Then, on one notable occasion, our assistant chief aerodynamicist Richard ("Dick") Jones[6] asked me to drop in for a chat—as I thought concerning another of my projects—he didn't beat about the bush and came right to the point: Had I published any work on gravitation in the United States prior to joining the company? As this was so, I answered in the affirmative. A broad grin of relief came over his face and after thanking me, he related the following extraordinary story.

It appeared that at a recent meeting with several visiting four-star American generals—who had declared interest in the SR53 and SR177 aircraft projects—both Dick and his chief found themselves paired off into separate groups as they toured the factory.

They soon detected a strong feeling that the visitors seemed somewhat distant and were not quite so intently interested in the aircraft as one would have thought, rather they were asking random, general 'fringe' questions with more than a hint or two about *gravity!* After a while, the two were able to arrange a moment to confer and so—realising they were both getting similar 'signals'—decided to play along with the situation. After several hours of this game of call-my-bluff, the charade came to an explosive and abrupt end over lunch, when the General sitting opposite Dick—while still in mid-sentence concerning the aircraft—suddenly leaned forward and, red in the face and with half-throttled hiss said, "Damn it, man, give us a break, what are you getting for your negative mass ratios?" Dick could only guess what the name of the game was and, taking a stab in the dark, proffered the first mischievous retort that came into his head. Apparently he acted his part suitably convincingly, for after furtively looking over his shoulder, he leaned across the table and whispered "about 0.05 or thereabouts," whereupon after a cliffhanger pause the General beamed gratefully, slumped back into his chair and then, as if nothing untoward had happened, resumed sudden rekindled interest in the aircraft!

For months after, Dick and his chief were at a loss to know what the affair was all about, save for the fact that the American visitors were extremely interested in matters gravitational. Dick Jones's conclusion was to me a little embarrassing, for after all I was a comparatively 'new boy' at the firm. He figured the Americans, on becoming aware of my active work on gravity, had quite erroneously assumed that it reflected the interests of my employers, who in fact knew very little about these private endeavours!

A Case of Deja Vu

One afternoon, some twenty-three years later in 1979 and several years after I had left the British Hovercraft Corp., I received a telephone call from a colleague still employed there, saying they had been entertaining some Russian visitors—both civilian and military—who *during lunch* had been asking inquisitive questions about my book, *Piece for a Jigsaw*; had I any remaining copies available? If so, they would send a man round to collect them. The following day I received yet another call; the Russian visitors were catching the next boat from Ryde and if possible they would like to have a *few more copies.*

The Russians got their additional copies and made their boat connection. My friend at the office was pleased and in a quizzical sort of way so was I. Yet how could I even begin to tell him of the strange deja vu episode in which he had just been an unwitting player? This sort of thing happens not only in novels and films, it happens in real life industry, and it happened to me.

Perhaps I will be forgiven if I admit to now and again having indulged myself the mischievous idea as to how much the world's future may have hung on R.S.J.'s chance retort!

Footnote

The foregoing can be summarised by saying that in the context of the developments we shall examine later on, there exists among every society a faction which appears to be almost programmed to strongly resist new ideas, but once a new idea becomes 'respectable'—usually by sheer weight of popular opinion—the same faction often covetously adopts the idea as their own, not always with the best interest of those they are supposed to serve.

There is unquestionably an increasing awareness of the need to curb global warming, coincident with the concern for the diminishing energy reserves. The signs are that these are both manifestations of one basic need, which will reach a meeting point sometime after the year 2000.

This work is an attempt to offer evidence towards a possible overall solution, which if correct will coincide with that same date. In the next chapter we shall be examining evidence for verification of that time and mankind's questionable fitness to have such a prize.

[1] More fully presented in a little-heard-of but exciting book titled *View from Orbit 2,*" T. Werner Laurie, 1956.

Note: In the original article published in 1954, Avenel anticipated the acceptance of the theory in *the year 2000*!

[2] This has never before been publicly aired during the past forty-two years.

[3] This has no relevance to the fusion experiment by Prof. Fleischman and Pons.

[4] See Appendix 2

[5] Appendix 1

[6] The late Richard Stanton Jones (eventual Managing Director of the British Hovercraft Corporation)

2
SETTING THE STAGE
Time, Pollution and Deadline 2000

A Kind of Time Machine

It is interesting to note that the years after 2000 appear to have some very special significance, for apart from being the geologist's estimated deadline when the world's fossil resources will begin to get seriously depleted, other predictions—not of the astrological kind—have been made which coincidentally imply that very same era. These predictions are of the kind by which scientists can quite accurately forecast the likelihood of forthcoming events in the natural sciences by observing past changes in terms of time. For instance, it has long been known by ecologists that the population biologic species, if given favourable environment, will increase *exponentially* with time. That is, the population will double repeatedly at roughly equal time intervals.

Both the world's automobile 'population' and the civil aviation scheduled flights are *doubling* every 10 years. The world's electric power capacity is growing at over 8 per cent per year and *doubling* over 8.7 years, while the human population is now doubling in less than 34 years. All of these trends plotted against time produce the same type of curve on a graph, called an exponential curve; that is, a curve which begins quite flat, then rapidly develops into a circular shape before rising almost vertically, as in Fig. 1.2. Normally this would have been interesting but of little significance, but for the fact that so many other 'event' curves develop in exactly the same manner.

Researchers have plotted man's speed of mobility against years, i.e., from the day he ran as a cave man, to the period when he first rode a horse, on to the time he copied a rolling boulder and constructed a wheel, then a canoe, on to a sailing boat, a car, locomotive, aircraft, spacecraft—the resulting curve is always the same! They plotted available power against time again with the same result. In this they discovered that the average modern housewife had at her fingertips—with the aid of the washing machine, kitchen equipment, radio, television and car, etc.—the equivalent of a baron of old living in a castle complete with horses and no less than thirty serfs at

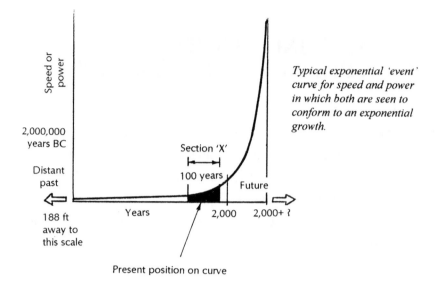

Typical exponential 'event' curve for speed and power in which both are seen to conform to an exponential growth.

Speed or power

2,000,000 years BC

Distant past

Section 'X'

100 years

Future

188 ft away to this scale

Years

2,000

2,000+ ?

Present position on curve

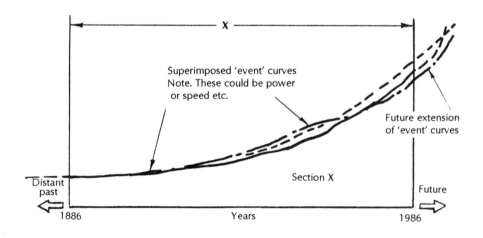

X

Superimposed 'event' curves
Note. These could be power or speed etc.

Future extension of 'event' curves

Distant past

Section X

Future

1886

Years

1986

Enlarged section showing how several superimposed 'event'
curves combine to form one exponential curve

Fig 1.2. Exponential Prediction of 'Event' Curves.

his command! When these and other 'event' curves were superimposed over the same time scale, their combined profiles also supported the exponential curve.

It can be argued that on this basis there is no reason to assume that such development will cease; if so—at the rate shown—the curve begins to approach the vertical portion sometime after the year *2000*. Further development of the curves indicate that by the year 2020, the same housewife could have at her fingertips the equivalent of several times that of the present level, and the future housewife of the 2070s could *theoretically* have at her disposal the entire output of a medium-sized power station, while space transit velocities at near the speed of light will be possible. This is exactly what the author's energy and propulsion theory predicts, but I did not believe it could happen quite so soon.

As if this is not incredible enough, later on we shall examine evidence for the process of Mat-Demat (Materialisation and Dematerialisation) and how theory suggests that not only is it logical that such phenomena may spontaneously occur, but it is quite feasible that in the future we may be able to produce such material effects electronically, thus vindicating the teleportation technique employed in the Star Trek series.

How long it will take for the last bit of the curve to become *truly vertical* is best left to speculation; suffice perhaps to say here that if and when mankind reaches that station, his knowledge would be total or—without wishing to offend or appear sanctimonious—he may become one with God (more graphically represented in Chapter 5).

But some scientists argue that such exponential growth—either biological or industrial—may be only a temporary phenomenon because the earth itself could not tolerate more than a few tens of these doublings of any biological or industrial event. They suggest that there can only be three possible 'event' trends: (1) The curve may—as in the instance of water power—level off and stabilise at a maximum; (2) it may go on rising and after passing a maximum, decline and then stabilise at an intermediate level at which it could be sustained; or (3) it may decline to zero again and become extinct [Fig. 2.2 (a)].

This argument is quite sound within the constraints of mankind being confined to this planet, but what if he extends his 'event' to include space? For example, the world's population growth may decline because we wish it, but on the other hand, if mankind begins to populate other planets, such increase would be of little consequence anyhow. Moreover it is interesting to note, that although the event curves generally follow the exponential rule, they may also include mini-type (1) curves, i.e., first rising exponentially, then levelling off slightly, before climbing once again. Of course, such mini-curves represent mere hiccups when viewed in terms of the total

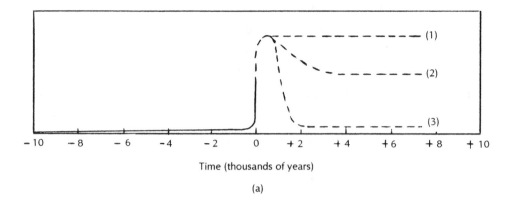

Time (thousands of years)

(a)

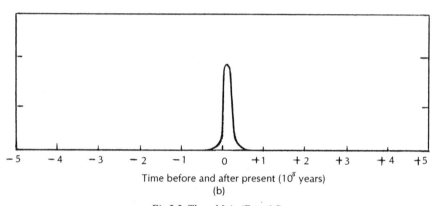

Time before and after present (10^5 years)

(b)

Fig 2.2. Three Main 'Event' Curves.

event line and would be completely invisible to the scale of that shown in Fig. 2.2 (b). This is exactly what we would expect of a *successful* culture, but for a failing one there would be a time when one of these hiccups became an extremely bad cough, degenerating into a type (3) curve descending to zero, whereupon after—perhaps eons—the whole process *very* slowly begins all over again.

To a degree, this method of prediction or guide can be regarded as a real technological forecasting device, a kind of time machine in the broadest sense, and over the past few decades the author has lived with knowledge of it and seen it vindicated time and again with every passing year. In my own world of aerospace I have observed speeds rising higher and higher from hundreds of miles per hour to thousands. I have seen rockets and space flight develop from the time when as a youngster, to me these were merely fantasies. Similar advances in other disciplines have kept an even pace; in medicine, metallurgy, meteorology, electronics, automation, lasers and computers, etc. Calculations which used to take me hours on a slide rule, a pocket calculator now does in seconds. In 1946 the E.N.I.A.C. computer occupied a room over thirty feet wide and fifty feet long and weighed some 20 tons! Today the same amount of information it could process can be handled by one silicon chip less than a quarter of an inch square. Indeed, although several decades ago it was impossible to understand just how the rate of development indicated by the exponential curve could be envisaged, today that is not such a difficult task, when we consider that the above computers are now taking part in the design of better and faster super-computers.

Perhaps being a very enthusiastic aerospace engineer and UFO researcher, it was natural for me in the 1950s to invent a convenient analogy in order to convey a visual interpretation to an audience. In this context, it seems appropriate to use it again here.

It is interesting and perhaps pertinent to this study, to view our current technological status through the eyes of visiting space explorers from a technologically advanced culture. Being familiar with the exponential curve law, one glance would suffice to reveal to them that mankind was now on the way to his rightful place among the stars . . . or Armageddon as he so chooses.

How often have we come across the remark "it was ahead of its time." "It," meaning of course someone's ideas that in the past have simply remained just that, ideas, or in some instances may have been translated into some kind of action, be it in terms of technological hardware, or purely ideological or philosophical persuasions. The ideas may have been sound enough, but somehow they did not relate to the prevailing situation.

Extraordinarily enough, among science fiction writers the high probability 'hit rates' are rather astonishing, the classic example of course being by Jonathan Swift—author of *Gulliver's Travels* (1726)—who described the two satellites of the planet Mars, 'Phobos' and 'Deimos'—including their size and orbits—long *before* they were discovered telescopically! Indeed it was the establishment's magnanimous way of paying Swift homage by officially recognising these names.

It is also true to say that practically every *new* idea can usually be shown to be a re-emergence of a similar, or even identical idea formulated long ago. During my lifetime I have met and heard of many inventors, who having worked on a new concept for many years—sincerely believing originality—have been later discouraged to discover that they were not the first. In fact it has happened so often to me personally that I have developed a philosophical pet belief and associated remark: "There's nothing new under the sun." For what it's worth it seems to me that we *collect* existing ideas—*substance* from the universe, modifing or reshaping them to accommodate our whims and putting them out as *new* ideas; and in that sense only, they are.

In the 1930s, when Sir Frank Whittle was working on his embryonic jet engine, he was totally unaware that in Germany, Hans von Chain was working on an identical idea and for that matter neither did Von Chain know about Whittle.

Whittle was short and of a humble background, Von Chain was tall and came from very wealthy stock. They both took on unimpressed aero-engine manufacturers and indifferent Air Ministries and won. On *both* their first successful prototype flights, the test pilots were unable to reach the hoped-for speed due to the fact that the under-carriages of *both* aircraft *failed to retract!*

Long before Sir Christopher Cockerell invented the peripheral jet system for hovercraft, the phenomenon of ground effect was well known to aircraft designers. In fact, every aeroplane or helicopter that ever flew experiences a degree of so-called cushion effect when operating near the ground or surface, whatever that may be. Many inventors tried to exploit the idea, some of them well known, many unknown. One of these, although historically well known, his interest in hovercraft, or surface skimmers, certainly was not. 'He' being none other than Lawrence of Arabia, who with the help of an aircraft technician, built and tested a piloted version!

Early designers independently pursued similar paths, to produce a vehicle which would operate off-road and skim along at significant air gaps (height above interface). But with the introduction of Cockerell's peripheral jet system and consequent increased research, it soon became evident that the off-road, over-hedges concept was going to be neglected due to an inherent limitation in the idea. That is, the higher a surface effect machine hovers, the more power it requires, until a point is reached where the craft is operating outside surface effect benefit and therefore requires much the same power loading as the helicopter. In other words, the closer to the surface a hovercraft 'flies,' the more efficient it becomes. But an additional problem other than efficiency is met at increased height, that of inherent instability. This becomes a serious problem when the machine operates at heights divided by the craft effective dia. (h/de) equaling more than .2. Therefore, these two reasons alone were sufficient to encourage designers to look for alternative solutions—including the present author.

Now it is obvious if you operate such a machine too close to a rough surface it is likely to sustain damage and that is exactly what happened to the first Saunders Roe SN1 hovercraft. However, as shown in my book *The AT Factor* several years *before* Cockerell's invention, my eldest son solved the same problem with a small experimental model by wrapping a strip of flexible tape around the bottom periphery. After offering some well-meaning tentative compliments, I, in my questionable wisdom, pointed out to him, "But you can't do that, son, it's cheating." That simple piece of tape was probably the world's first single membrane flexible skirt and *several years* ahead of its time. After some eighteen months' design, rebuild and further SRN1 damage, it was eventually introduced by Cockerell as a new idea. More significantly, without those skirts the hovercraft as it became could never have advanced far beyond the stage of yet another interesting idea, with or without his annular jet.

Another not so well-known example of the 'time advance' variety relates to the author's work in that there is good reason to believe that had the eminent English experimental physicist Michael Faraday (founder of the science of electromagnetism in the 19th century) lived a little longer to continue his last researches, he would more than likely have identified the same primary energy source described in this book. In which case, we are left to ponder: if after 119 years, one of the chief objectives of this book is in effect to say, not so much *how* shall we do this thing, but *should* we, what might the world have been like today?

Primary Energy

But, if sceptics would say such 'foresight' is a natural function of the mind being influenced by *existing* stimuli, how would they explain the fact that centuries ago people visualised an aspect of nature which only now is beginning to be suspected by a few modern thinkers? I refer in this particular instance to Cosmic Primary Energy.

The belief in a primary energy goes back into recorded history. In the Sanskrit writings it is called 'Vril,' the yogi calls it 'prana' and contemporary scientists of the early 19th century called it the 'ether.' Although it was discarded—due to the negative results of the Michelson-Morley experiment in 1897—some scientists are now becoming reconciled to some form of an ether or energy field.

Indeed, it will be shown later that what we call 'empty' space *is* something, that it and matter are synonymous—in a word, they are different in degree, not in kind. It is the stuff or 'state' from which all matter (secondary energy) is formed. It is the 'essence' which *was* before matter *is*. It is *not* that which we use when we cause matter to react on matter—as in Newtonian physics—indeed it is the 'state' to which matter ultimately returns; we commonly call it *space*. The author considers it is more correct to call it the *primary* energy state, but later on we shall see why it may be more technically correct to employ the abridged term 'Comat.'

In accepting the above we are in good company, for it was Sir Oliver Lodge, the famous nineteenth-century physicist, who gave the simile that "the electron and space could be likened to a knot on a piece of string." Incredibly, orthodox science hasn't even begun to recognise the fact, let alone consider the possibility of using such power, which—as I intend to show—could solve all our energy, transport and pollution problems. Even so, if it *is* true that there is a time and place for everything, in the light of the following can it *really* be now?

Inadequate Methodology, Pollution, Greed and Entrenched Politics

We cannot continue to accept the fact that every day over *thirty million tons* of sulphur are emitted from industrial plants throughout east and west Europe alone. It is claimed that the bill required to eradicate this problem would be of the order of fifteen billion pounds per year and we cannot afford it. But a tiny fraction of this colossal sum alone spent on research of the kind presently discussed could pay enormous dividends. While on the other hand, many are they who—while claiming to have the world's interests at heart—are taking part in its wholesale decimation, and this not just for a living, but gluttonous enormous profits. At this time pollution is now being exported to the newly developing countries by large corporations eager to escape from the tightening pollution controls in their own countries, thereby taking blatant and indifferent advantage of the needs of these developing communities, known as pollution *'havens'* like Cubatio in southeast Brazil. Here an open invitation was given to industry—both local and international—to locate its 'dirtier' plants in the valley. This concentration of industry with relatively low pollution control has earned Cubatio the title of the dirtiest town on earth, and many of its people pay for the lack of control with their lives. Chemical waste build-up in some ditches is so severe that it has ignited into a fireball! It is now public knowledge that in the case of the Union Carbide pesticide factory at Bopal in central India in 1984, it was inadequate control and unregulated housing close to the plant which added to the impact of that disaster. The leakage of toxic chemicals claimed around two and a half thousand lives and left thousands of people injured and suffering from long-term physical defects.

Concerning this, one writer said, "These examples of the export of pollution from developed countries to the newly industrialised ones highlights environmental inequalities, indeed environmental inequality is in itself a form of social inequality. It is created by a system of dominance by advanced industrial nations, over the dependent newly industrialised ones." On a similar note, in 1987 Tate and Lyle's former research chief Prof. Chuck Vlitoss said, "I think it a very important question to ask ourselves, where will all the chemicals that we are used to today come from in the future, when we run out of petrochemicals, where will they come from? You can't produce chemicals from nuclear energy, it just doesn't happen, you have to have an alternative energy source and the only one I can think of at the moment is in fact plant material, whether that be cellulose or sugar, trees or sugar crops doesn't matter, *but we have to begin today to do that research.*"

Desecration of a Planet

Nearly all airborne pollution—such as the familiar smog—is due to energy consumption of one kind or another, by automobiles, power plants and industrial centres, etc. The pollutants consist chiefly of carbon monoxide, particulates, sulphur oxides, hydrocarbons and nitrogen oxides, the proportion of which varies with the polluter. So that the generation of electric power is the major source of sulphur oxides, whereas the motorcar produces more carbon monoxide, hydrocarbons and nitrogen oxide.

It is not possible to determine the importance of these pollutants from their gross weight because they have very different effects. Some, such as carbon monoxide, affect health in even minute concentration; others like particulates largely add to the cleaning bills.

At present the generation of electric power depends chiefly on fossil fuels and sulphur oxides come from the sulphur impurities in these fuels. The burning of these fuels also converts considerable amounts of carbon to carbon dioxide. Although this familiar gas is not a pollutant in the ordinary sense, its steady increase in the atmosphere is cause for concern. Carbon dioxide is largely transparent to the incoming short-wave solar radiation, but it reflects the longer-wave radiation by which the earth's

Just one of the fourteen enormous tanks in a large oil tanker. The small figures in the background offer a visual measure of the sheer size of these colossus. There are many such ships travelling round the world every day.

Fuel	Years @ 5% growth rate
Oil	55
N.Gas	75
Coal	155
Uranium	95

Fig 3.2. General Scenario (for 1987) of the world's fuel reserves. (a) top, (b) bottom

heat is radiated outward, thereby causing the temperature of the earth to rise, thus producing the well-known greenhouse effect. If this continues, estimates predict a possible rise in the earth's temperature of 4[degrees]C by the year 2050. This would prove disastrous for the world's agriculture *and cause massive flooding.* At present about 6 billion tons of carbon dioxide are being added to the earth's atmosphere every year, increasing its carbon dioxide content by over 0.5 per cent per year. By the year 2000 the increase could be as much as 25 per cent. Some experts believe that if we go on burning coal at an ever-increasing rate, pollution would be raised to a *totally unacceptable level by 2020.*

At present, our understanding of the atmosphere is insufficient to predict the eventual changes in climate that might be produced by the increase and by the related increase in water vapour and dust content, but even small changes in the average temperature could have catastrophic effects.

Some of these pollutants are responsible for the so-called 'acid rain' much in the news lately and considerable investigation is underway to try and determine if in fact this is responsible for the alarming increase in diseased trees.

At Kitty Hawk in 1903 the Wright brothers flew the first man-lifting, engine-powered aeroplane. Sixty-six years later—less than one lifetime—two American astronauts, Armstrong and Aldrin, landed on the moon. To have only a bare indication of the enormous technological strides that this near miracle entailed is one thing, it is quite another to have a closer acquaintance with it. But the associated romance and glamour which surrounds aerospace activity is inclined to somewhat overshadow other equally technological marvels. Indeed it is due to the fact that industry kept up an even pace that such developments were able to occur.

Even so, with this in mind, it is not easy to accept the fact that as a technological society we are still quite young. A German physicist points out that according to the exponential rates, previously discussed, our 'technological age' is barely 200 years old and he estimates that *under ideal circumstances* the life of a 'technological society' is around 2000 years! After which he maintains, such a society would have literally used itself up. In a word it would have exhausted not only its energy resources, but its raw metals as well. Note that here again the assumption is made that man continues his existence and 'technological life' only on the Earth. At only a tenth of our estimated life, our society already faces an immediate—50 to 100 years—shortage of fossil fuels and already we are facing a possible strain on our raw material resources, although this is not at present such a serious problem. Great Britain, for instance, uses only 3.5 per cent (300 peta jules) of fossil fuel energy for lubricants, petrochemicals and plastics, etc. But even for this limited use, it cannot last forever! [Fig.3.2 (a) and (b).]

Embracing as it does these and many other main issues concerning our lives and future on this planet, there can be little doubt that the subject matter in this book will prove to be highly controversial, if not contentious, depending on your attitude and station and—as already stated—it is offered here to anyone who has the capacity to reason and judge for themselves.

In some respects it may be very difficult for some lay people—and experience has shown, for many technically trained ones as well—to be able to take a detached view of mankind's technological achievements, but right from the beginning of this analysis we have to achieve that status. Beyond question, the technological strides made over the past few decades is a remarkable achievement of an extremely high order and even a brief acquaintance with a jumbo jet and the Wright brothers' canvas-and-wood biplane Kitty Hawk bears testimony to that. However it is also true that many are inclined to the view—as were most of their forbears—that we have 'done it all' and there is little room for improvement. It will become evident later on that the arrogance of those who are inclined to believe that man's many diverse mechanical

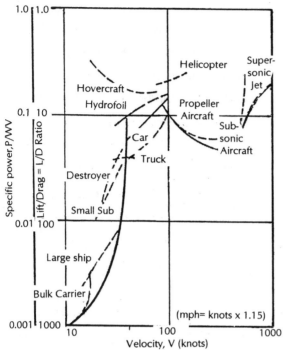

Fig 4.2. *Specific power plotted against velocity shows how existing vehicles compare.*

solutions reflect his engineering supremacy is self-deceptive, for in truth such compromising efforts constitute his admission of defeat. As yet he knows no other way.

In 1950, two engineers by the names of G. Gabrielli and Th. von Karman published a plot in a technical article entitled "What Price Speed?" In the paper the authors plotted the specific power required for the propulsion of existing vehicles [Fig. 4.2]. The plot is just as valid today and it is included here—for those who wish it—merely to offer visual evidence of the design compromise. Indeed it can be stated, that if this was not so, we would have had just one method of transportation for *all* applications long ago. That time may now be very near, and I think it is already accepted by a great many thinking people that some of us will look back with incredulity to the days when—for instance—aircraft *had* to belt along terribly expensive runways at alarming speeds and angles in order to take off!

Cars, Trains, Boats and Planes: A Compromise

Diversity born of compromise is illustrated only too well in our transportation systems. In fact most of our present day manufactured commodities from cars to washing machines, aircraft to nuclear power stations, all are based on this system of design compromise.

Again, it is interesting to consider some of our present transportation systems through the eyes of our technologically advanced space visitors. A brief look at this diversity, ranging from cars, ships, air and space craft, etc., would reveal instantly to them our as yet—albeit necessary—relatively kindergarten technological status. This appraisal is not borne of any inverted bias on my part; after studying the statistics the reader—and for that matter any engineer worth his salt—would agree as much, for all of these various methods represent our desperate attempts at engineering compromise in order to scrape just a little bit more efficiency from the various systems. Our space visitors would recognise that all of our transport vehicles are designed to give the best return for a specific task. Thus, the high speed jet propulsion system is the best for high speed aircraft, but little use for slower aircraft and vice versa. The slow ship is more efficient for carrying massive loads, while on the other hand it is better if the vessel can be lifted out of the water in order to reduce drag and then we get a hydrofoil or a hovercraft, but they achieve this by lifting less loads for a higher power consumption. A modern motor car is capable of travelling fast along a motorway, yet unless completely redesigned lacks the basic capability of lifting itself *vertically* over a one-foot-high brick wall; it requires a hovercraft or a helicopter to do that. Equally—and even more diverse—a hovercraft or a helicopter cannot travel under the water, or out into space, we require a submarine vessel or spacecraft to do that. A helicopter can take off and land vertically for a given installed power, yet for aerodynamic reasons cannot fly as fast as a similarly loaded conventional aircraft.

Of course there are specialised exceptions to this general trend, but the lack of their numbers illustrates only too well the lack of success. As illustrated in Fig. 5.2 (a) and (b), associated running costs are involved, but again we come face to face with the need to compromise. In fact, here and there in this book we shall see evidence of this tendency and it will be shown later that this limitation in our technology is but a measure of our present lack of knowledge about the formation of the material universe. It will be shown that not only is complexity and the need for diversity unnecessary, but high cost also.

It is anticipated that at this point some readers will instinctively and rather naturally object to the above observations. "How else," they will ask, "for don't we require all these things for their specific tasks?" The blunt answer is, "of course, given our *present level of knowledge,* precisely." To those I say, be patient, for by the end of this book you will probably have completely changed your mind.

Let there be no mistake, as an aerospace engineer I too marvel at the spectacle of a cathedral-sized machine called a spacecraft blasting off a launch pad to place equipment and a mere handful of people into earth orbit. Yet even lay people often instinctively feel there is something fundamentally wrong about the sheer size of these colossuses, and for that matter—albeit privately—so do many engineers. Bearing this in mind, perhaps the reader will be brought closer to identifying with the author of my particular no-man's-land, which is the result of a peep into some future realities.

International Politics of the Year 2000 and Advanced Technology

Beyond any doubt one of the greatest problems to a realistic adaptation of a radically different energy and transport system is the enormous uncertainty of how

Step	Step Efficiency	Cumulative Efficiency
Production	96%	96%
Refining of crude oil	87	83.5
Transportation of gasoline	97	81
Thermal efficiency of engine	29	23.5
Mechanical efficiency of engine	71	16.7
Rolling efficiency	30	5

(a)

The Energy-System Efficiency of the Automobile

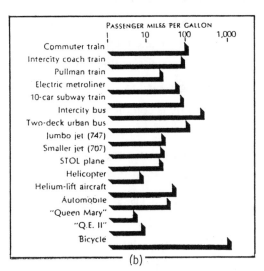

(b)

Fig 5.2. The propulsion efficiency of various transport
systems. The bicycle is shown on an equivalent energy basis.

our society could cope with it, not only in our home country but at international level. We are reminded daily by the media of the suspicion, mistrust and downright inability of some of the world's population to forgive and forget past errors, largely germinated by fears, greed and misinterpreted religious ideology of our forbears. But it is now becoming difficult indeed to see how—*in the light of our advancing modern technology*—we can continue to be anything other than forbearing to one another. For instance, gone eventually will be the days when counterintelligence departments spied, bugged and connived for a living. The following somewhat crude, but nonetheless valid example may help to establish the point.

Earlier in this chapter I mentioned the technological developments I have witnessed in my own time, from slide rules to supercomputers, etc., and I am reminded of this and other relevant issues even as I pen these words, due to the fact that I have on my desk a daily reminder in the form of a cheap digital watch—which was totally beyond my comprehension in those earlier days—chiefly manufactured of plastic, which keeps more reliable and accurate time than more expensive types I had when I used the aforestated slide rule. The difference being, not only does this one represent a technological breakthrough born of the space age, but is is now so cheap and common it was *given away* as a sales gimmick with an equally cheap pair of foreign Trainer shoes!

Now, with no wish to overstretch the credulity of the reader, I put forward the following hypothetical scenario—which of course is not intended literally—merely as an illustration concerning the ultimate uselessness of future intelligence surveillance. It is a typical scene which based on the rapidity of present exponential development—borne out in this particular instance by the fantastic strides currently made in 'nano' engineering—could occur sometime into the next century.[1]

It is nearing Christmas time and you are ambling around the supermarket store with your wife looking for a suitable 'little extra' to add to your ten-year-old son's presents. Your time and your pocket are short, as is your patience with your inability to find something interesting and with credible play value. Your wife draws your attention to something she had spotted earlier in the week, as being potentially amusing. The toy consists of a small flat-screen TV monitor, a small control console complete with stick operation and a printed card with transparent press-out dimples. Despite the all too obvious packaging title you find yourself minutely examining the small objects beneath their transparent covers; they prove to be nothing more than perfect replicas of the common housefly!

Among other things the instruction leaflet goes on to inform you that not only does this object comprise the most up to date super-microtechnology consisting of 'on-board' battery—rechargeable of course— micro-electric motor, solid state gyro/rate sensor and 'true to life' wing propulsion system, but a complete AVRT (audio/visual receiver/transmitter), etc.

At that particular time in your not too distant future, you are amused but not overawed by this remarkable adaptation of modern technology to reproduce an exact working model of an RP (remotely piloted) housefly, but neither are you amused at the prospect of being surveyed in some of your more private moments! "Brilliantly clever what some of these foreigners get up to," you muse, "though the damned thing could cause mischief and should be banned." Noticing your hesitancy, your wife mischievously assures you, "They're all the rage in the States at the moment and selling like flies!"[2]

I emphasise, based solely on the technology now being developed, that this little RP fly *is* a possibility, although its eventual manifestation is most unlikely. But the narrative may serve as an illustration of a future technology which would determine that we could no longer have *industrial or political secrets from one another!*

I repeat, there is no need to try to interpret this analogy literally, although from it we might gain acceptance of the possibility of an emergent technology presently beyond most of our dreams, as was the modern pocket calculator beyond mine, which will certainly be just as effective as would be the author's hypothetical fly surveillance system. Which incidentally in the remote eventuality of it being detected, fly sprays would be absolutely useless! Neither does the time-expired argument, "to every technological spying device there is always going to be a technological countermeasure" continue to be true, as the success of high flying reconnaissance satellites has shown.

So whether we like it or not, the World as we know it is going to change. In order to cope with the technological ramifications, we *will* have to show a greater degree of forbearance toward one another and what is more, clearly we haven't much time to achieve that.

The remainder of this book is devoted to my—no doubt often inadequate—efforts and, for what they are worth, my attempts to interpret some of the information concerning gravity, energy and the nature of the Universe which has come my way—sometimes in the most remarkable circumstances. It is my sincere hope it may succeed in helping a little.

[Some of the material in the following chapter forms an abridged extract from *Piece for a Jigsaw* published in 1964, which due to extraordinary circumstances was withdrawn from circulation. It is reassuring that during the whole of the ensuing time, there has been no cause to change the general view.]

---

[1] At the time of writing researchers have successfully developed a tiny steam engine, the piston of which comprises a few molecules and functions on a nano droplet of steam!

[2] I used to offer this simile at lectures years ago, certainly long before the advent of the X Files.

3

GRAVITY
The ether, 'the strong force,' matter and energy

"He that removeth weight doeth as much advantage motion as
he that addeth wings." John Pym (1584-1643)

For many years I have never ceased to be moved by the farsighted validity of
this remarkable statement made by John Pym *over two hundred years* before the
Wright brothers flew! During the progression from aeroplanes to rockets, gravity,
space field propulsion and energy, it has never been far from my mind.

My reasoning in earlier days was—and for that matter still is—that mankind's
discovery and eventual employment of magnetism was originally aided by a natural
phenomenon called the lodestone, which he couldn't understand. Similarly with static
electricity, he had only to stroke a cat to observe another phenomenon which equally
he couldn't understand. But both of these phenomena were repeatable and they could
be isolated or shielded, measured and eventually understood. However gravity is not
only a phenomenon which equally we don't understand, but normally it cannot be
shielded for repeater experiments in the laboratory and even if it could be, we are
penalised by the relatively exceeding weakness of it.

Seance Room or Laboratory?

On the other hand even a superficial enquiry into levitation of matter by paranormal means immediately suggests isolation or shielding of some kind. For instance,
among other things serious investigators will discover is that when objects are paranormally levitated they appear to lose *inertial resistance* in *all* planes, i.e., laterally as
well as in the vertical sense, which is exactly in accordance with Einstein's predictions. Thus it is in the seance room we are more likely to find repeater experiments in
the pursuance of gravity, which can pay great dividends in the formation of ideas, as

I have found.

On a more philosophical note perhaps, it is in the correct order of things that an understanding of the nature of gravity cannot, must not, be prematurely acquired, for some very obvious reasons when you think about it. Since this is the most important of the three apparently different phenomena, it seems fitting that the acquisition of a 'laboratory' tool for gravity research should first require a degree of open-mindedness among researchers.[1] Not that I am suggesting that physicists should go stampeding the seance room door, it requires a great deal more patience than that. But I am restating *they are extremely handicapped if they deny paranormal effects.*

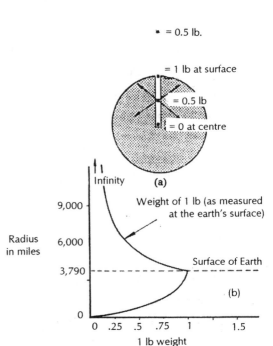

Fig 1.3. Gravitational 'attraction' is a function of masses and the radius between the mass centers.

Gravity

What exactly do we know about gravity? Beyond measuring its effects, very little indeed. We do know that it is the all-pervading aspect of nature which dominates all the others. Without gravity, stars wouldn't form and the universe would have no meaning. We do know that it not only holds down the slates on the roof over our heads, but it also holds us down on the seat of our chairs, there is no apparent screening or shutting off effect between them. What is more, beneath the floor on which we are sitting there is another person similarly held down on *their* chair and *their* floor and should our building be a skyscraper there could be many people similarly held down on floors by this enigmatic force, but still there has been no noticeable diminution of gravity effects at any level. We shall see later that in fact what diminution there is, is so small as to defy detection by ordinary means.

While it is true to say that due to the inverse square law, the gravitational 'pull' exerted by a planet on a smaller body increases the nearer the body gets to the planet, it is not true to say that the 'pull' on the body will continue to increase if it could descend a great pit towards the centre of the Earth [Fig. 1.3 (a)]. For it will be apparent that the moment the body goes below the surface, then the surrounding mass is exerting a force from all directions on it, until we can visualise a condition when on reaching the centre of the Earth, the 'pull' or weight of the body would be zero, according to the graph (b) or nearly so.

The Inferiority of Gravity

Of the known forces in nature, gravity is by far the weakest. For instance according to classical Newtonian physics two protons of 1.66×10^{-24} gm weight can at lcm distance, by virtue of their masses, 'attract' each other with a force of $(1.66 \times 10^{-24}) \times 66.8 \times 10^{-9}$ dynes $= 18.10^{-56}$ dynes. At the same time they exert an *electrical* 'repulsion' of $(4.774 \times 10^{-10})^2$ dynes $= 22 \times 10^{-20}$ dynes. The electrical force is therefore roughly a sextillion (10^{36}) times as large and of opposite sign to the gravitational 'pull.' If like most people the reader finds it difficult to visualise such quantities, 10^{36} looks like this:

10,000,000,000,000,000,000,000,000,000,000,000,000!

It is chiefly due to this relatively inferior strength of gravity that research has been so seriously inhibited. In a word, if they exist, it is almost impossible to detect gravitational waves and it is logical this is where research into so-called anti-gravity must begin. A great deal of work has been done by a very few dedicated pioneers, so far with little known success.

Combating Gravity and a Directive for the Future

Balloons, aircraft and all air supported vehicles (ASVs), *appear* to overcome gravity, but of course we know they do not.

With the conventional aeroplane, no matter how sophisticated it may become, man is combating gravity with the aid of aeronautical stilts, no more, no less. He rises from the bottom of the aerial sea he calls his atmosphere, much as do the fish in the oceans; neither they nor he are one jot free of gravity. The rocket is not much different. The principle is one of refinement, not of kind.

The aeroplane attains lift as a reaction to differential air pressures and/or displacement of a mass of air over its wings, while is is well known that the rocket attains 'lift' or thrust as a reaction to a rearward ejection of a mass of gas; fundamentally the principle is much the same. The chief difference lies in the fact that the rocket is capable of taking some of this gas, or 'working fluid,' outside the atmospheric envelope, where it can go on functioning in free space.

We observe how rocket vehicles seemingly oppose the force of gravity by piling on an enor-

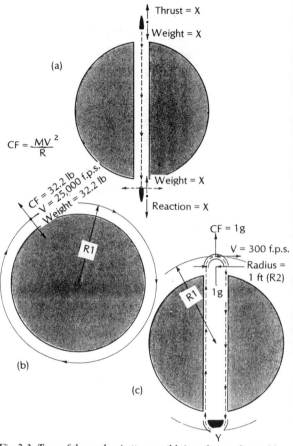

Fig 2.3. Two of the author's 'impossible' analogies derived by natural laws which nevertheless illustrate that there must be a more effective way of combating gravity.

mous velocity, which will either be fast enough to send the vehicle coasting outwards against gravity, or if directed parallel to the earth's surface, create sufficient centrifugal force to exactly match the gravitational 'pull'; and then it is said, the vehicle is in orbit [Fig. 2.3 (a)]. In any event, gravity is being fought, with man as the duelist, having not the slightest idea just what it is he is fighting. True, we are all familiar with the limitations of the rocket, but are there any grounds at all for considering that pure reaction machines may not be utilising their energy in the most efficient manner? Well, from the thermodynamic point of view, the efficiency of the modern rocket is rather good, and in this respect the thermodynamicist has done incredibly well. And it is obvious that as far as the available kinetic energy stored in the thrust jet is concerned, the engines have been developed to their maximum potential. But suppose an argument could be offered which although mechanically unrealisable, would nevertheless indicate that every reaction motor is *theoretically* capable of producing exactly twice its power? Then we might conceivably feel justified in saying to the rocket engineer, "You're doing an admirable job, but are you sure you are using the available energy in the most profitable way?" Then naturally he would ask for your reasons and having offered them to him, he might be offended by the simple nature of the argument, that is, if he accepted the analogy seriously.

The author has two such arguments to offer, but I would stress to the reader they cannot be interpreted literally, anymore than one could dream of building a bridge higher than the tallest mountain and circumnavigating the earth where it would remain suspended at its centre of gravity without any physical support. Theoretically possible, but quite unrealisable.

The first analogy in Fig. 2.3 (b) is self-explanatory, in which a rocket is just supported by its jet efflux, weight being equal to thrust. But the exhaust gases are conducted along a shaft running through the centre of the earth which—allowing for no losses in the kinetic energy of the exhaust stream—is theoretically capable of supporting another rocket of equal weight situated diametrically opposite on the other side of the globe. The reaction on the base plate on this second rocket being equal and opposite to the reaction of the first, so that two rockets are theoretically capable of being supported by the blast of one—neglecting the loss of weight due to consumed fuel. In fact, this situation exists at every rocket lift-off, for the deflected exhaust is in fact 'pushing' the Earth in the opposite direction, but due to the relative gargantuan mass of the Earth any displacement is virtually indetectable.

The second analogy is equally annoying, for while true and serving to illustrate the point, it can never be realised in actuality; it too concerns centrifugal force. In Fig. 2.3 (c) we consider again the fundamental law which keeps a satellite in orbit; centrifugal acceleration of 1g (being a function of the vehicle's forward velocity) balances the acceleration due to gravity, i.e., 1g. It will be seen from the formula $CF = MV^2/R$ that in order to maintain a constant centrifugal acceleration of 1g, any reduction in the value of R must be accompanied by a proportionate reduction in the value of V. But obviously orbiting capsules are restricted to R_1 in the figure due to the size of the earth. If however we could reduce R_1 to the relatively very small radius of R_2 in Fig. 2.3 (c) and a revolving mass of only a few pounds be permitted to enter the pit running through the centre of the earth, and revolve in the duplicate system at position Y, then two such identical earth 'satellites' of equal mass to that in case (a) could be suspended above the surface of the earth for a fantastically reduced amount of kinetic energy, that is, if we neglect the time taken in transit of the mass from one side of the globe to the other. Just for the fun of it, this means, in the hypothetical case of an 'orbiting' weight of 32.2 lb. revolving round a lft radius arc and travelling at, say, 300 ft per sec, would yield a force of:

$$CF= \frac{MV^2}{R} = \frac{WV^2}{gR} = \frac{32.2 \times 300 \times 300}{32.2 \times 1} = 90,000 \text{ lb}$$

or a little over 40 tons! In other words, the above 'engine' under these conditions is theoretically capable of supporting *two* 40-ton spaceships and—allowing for no frictional losses in the system—once the revolving mass had been accelerated up to 300 feet per sec, it would go on for all time, just exactly the same as its earth orbiting brother.

But if it takes colossal boosters yielding millions of horsepower to accelerate orbital capsules of several tons' weight up to velocities in the order of 25,000 feet per sec, then there is no necessity for further recourse to calculation in order to show that the acceleration of a thirty-two-pound weight up to a velocity of a mere 300 feet per sec—as in the latter example—would take a correspondingly *microscopic* amount of energy to accomplish several times the work. If only we could dig that big hole!

Long before, and certainly ever since Galileo first demonstrated that all bodies irrespective of their size or mass 'fall' towards the earth with the same acceleration, men have dreamed and pondered on the mystical nature of gravity. What is this strange force which permeates all matter, what is its origin and further is there any evidence to indicate that there could be a contra-gravitational state? It so happens there is, although at first it isn't obvious until you think about it. Nature has always been generous in giving mankind helpful clues, indeed the whole of science is based on the fact that nature is the teacher, mankind the copyist. Classic example being the beautiful aerodynamics of the sycamore seed, observable long before Leonardo da Vinci dreamed of his embryonic helicopter, while no doubt a boulder rolling down a hillside had something to do with the invention of the first wheel. A floating log probably introduced mankind to the principle of the boat, and who can deny the sun was a ready-made lesson on the nature of heat. So logically, if it exists, there should be indications for a contra-gravitational state. It so happens there is, for every log that ever floated and every cloud that was ever formed, and for that matter, every lighter than air balloon that ever flew, bears testimony of it. A few words for lay readers may not be amiss.

In these, probably mankind's very first truly contra-gravitational lessons, it is well known that lift is caused by Archimedian-type displacement or buoyancy, but what is not fully appreciated is that the phenomenon is caused solely by the action of gravity exerting downward pressure on the water or air which in turn thrusts the boat or balloon up.

Similarly, winds around the Earth are also a byproduct of gravity. Therefore when a lighter than air—or more correctly a displacement vehicle—rises and is blown along by the wind, it is both suspended *and* propelled indirectly by gravity, for free![2] Gravity also lifts the vehicle off the ground thereby conveniently minimising surface friction. There is also very little aerodynamic drag due to the fact that the device moves *with* the air. And if the wind could be considerate to blow around the Earth in only one direction, we could take off and travel around it for nothing! Indeed, as I pen these words preparations are afoot for two competing balloon crews to do just that. However this method is both slow and *not* directional, but we shall see later the 'Comat' power—which is the *origin* of gravity—*is* directional and its latent power and associated velocities are vast. The main point at issue here is the fact that what nature demonstrates in one form usually implies that the phenomenon can be reproduced by other approaches and other means. In a word there is more than one way to climb a mountain and as we shall now see gravity will prove to be no exception to this general trend.

There can be little doubt that the late Sir Arthur Eddington was equally addressing the present when in 1933 he wrote, "We have turned a corner in the path of progress and our ignorance stands before us, appalling and insistent. There is something wrong with the present fundamental conception of physics and we do not know how to set it right."

The following observations of the author are by way of an honest appraisal, rather than a rebuff, but let us face it, almost daily the enormous strides made in physics take scientists deeper and deeper into an ever-darkening wood and somewhere within that wood lies the answer to gravity. Only a few years ago, physicists spoke of the electron as a particle, but then the thing behaved irrationally as a wave and clearly it couldn't be both—or could it? Thus the 'waveicle' was born. Perhaps in no clearer way can the situation be expressed than by paraphrasing Hoffman's[3] chicken and egg riddle.

Little boy: "Daddy, what came first, the electron or the wave?"

Daddy: ". . .eh, yes."

The late Sir James Jeans once made several pointed remarks with which such men as de Broglie, Max Planck, Einstein and Schroedinger associated themselves. They may be thus epitomised: "Thirty years ago we thought that we were heading towards an ultimate reality of a mechanical kind. Today there is a wide measure of agreement, which on the physical side of science, approached to unanimity, that the stream of knowledge is heading towards a non-mechanical reality. The universe begins to look more like a great thought than a great machine. Matter is derived from consciousness, not consciousness from matter. We ought to hail mind as the creator and governor of the realms of matter."

The Ether

On gravity, Sir Isaac Newton in his *Principia* said: "That there is some subtle spirit by the force and action of which, all movements of matter are determined," and again in his third letter to Bentley he said:

> It is inconceivable that inanimate brute matter should without the mediation of something else which is not material, operate upon and affect other matter, without mutual contact, as it must do if gravitation in the sense of Epicurus be essential and inherent in it. That gravity should be innate, inherent and essential to matter so that one body may act upon another at a distance, through a vacuum without the mediation of anything else by or through which their action may be conveyed from one to another, *is to me so great an absurdity that I believe no man, who has in philosophical matters a competent faculty of thinking, can fall into it. Gravity must be caused by some agent acting constantly according to certain laws, but whether this agent be material or immaterial I have left to the consideration of my readers.* [Author's italics]

On gravity, Michael Faraday said:

> As the coil is to the magnet, *so I believe the condenser may be to gravity,"* and again, "I have long held an opinion, almost amounting to conviction, in common, I believe with many other lovers of natural knowledge, that the various forms under which the forces of matter are made

manifest have *one common origin*; or in other words, are so directly related and mutually dependent, that they are *convertible* as it were, into *one another*, and possess equivalents of power in their action. In modern times, the proofs of their convertibility have been accumulated to a very considerable extent, and a commencement made of the determination of their equivalent forces.

And again,

This strong persuasion extended to the powers of light, and led, on a former occasion, to many exertions, having for their object the discovery of the direct relation of light and electricity, and their mutual action on bodies subject jointly to their power; but results to this time have been negative.
These ineffectual exertions, and many others which were never published could not remove my strong persuasion derived from philosophical considerations; and, therefore, I recently resumed the inquiry by experiment in a most strict and searching manner, and have at last succeeded in *magnetising* and *electrifying* a *ray of light*, and in illuminating a line of *magnetic force*. [Author's italics]

Sir Oliver Lodge, in his book, *Ether and Reality*, stated: "If waves setting out from the sun exist in space eight minutes before striking our eyes, there must be in space some medium in which they exist and which conveys them. Waves we cannot have lest they be waves in something."

In *Sidelights on Relativity*, Albert Einstein said: "According to the general theory of relativity, *space is endowed with physical qualities . . .space without ether is unthinkable.*" If everything is formed by modulations of this ether, all our present energy sources are derived from secondary forms of it; *why not go to the source?*

Interestingly, early scientists of the 19th century had intuitively accepted the existence of an ether until the famous Michelson-Morley experiment in 1887 which failed to find evidence for it. However, since then more and more physicists are beginning to reconcile the existence of a modified ether which could in fact explain the apparent negative findings, which after all related solely to the propagation of light.

The tendency to over-compromise is not only restricted to our mechanistic or physical developments, it can be found in the area of theoretical physics as well. Indeed some would have it that this is the seat of all compromise. For instance it is now generally accepted that gravitational, magnetic and electrostatic forces are different phenomena; in this and following chapters I will be offering my reasons for suggesting they are not.

To say that it's difficult to imagine an all-pervading ether which permeates all space and is the seat of gravity, magnetism and electric fields is something of an understatement, for it's difficult enough to imagine an atmosphere yet we know it's there. But do we *really* have much of a conceptual feel for the atmosphere? Here is a simple little test to find out.

Hold your hand in front of you, *slowly* move it from side to side, flex your fingers; feel anything, resistance of any kind? Not one bit, yet at sea level the palm of an average-sized human hand is being subjected to a pressure of *over four hundred pounds!* But due to the fact that the back of the hand experiences this same pressure, all feeling of resistance is cancelled out. In a word if by any means the atmospheric pressure on one side of your hand could be instantly removed, in all likelihood your arm would be torn off! Not a very nice thought, but it may help to illustrate that seemingly nothing is in fact something very potent indeed.

A more convincing experiment does require special equipment, but a description may be sufficiently convincing. Well known to engineers is the fact that if two pieces of metal are ground to a high degree of smoothness then pushed together, they will immediately appear to be 'glued' together with such force they can only be separated by sliding them apart. This phenomenon is brought about due to the fact that most of the air between the blocks has been displaced, leaving only the surrounding air to push them together. Should the faces of the blocks be, say, one inch square, and the ambient atmospheric pressure approximately fifteen pounds per square inch, then the blocks have a total 'gluing' adhesion of some thirty pounds force acting on them.

Of course, with all material objects subjected to this same degree of atmospheric pressure, we still find it difficult to sense it. How much more so than with a totally imperceptive ether? Well normally yes, save in the case where the object happens to be a simple bar magnet.

With that in mind, let us repeat the exercise as before; that is, we move a magnet slowly from side to side, still we feel nothing, but now if we produce a piece of iron or steel and bring it and the magnet together, they are similarly 'glued' together requiring considerable force to prise them apart. Thus we have two identical phenomena in which two blocks of metal are 'glued' together by nothing other than an air pressure differential on the one hand and another two blocks 'glued' together by nothing other than magnetism on the other. Is it any more unreasonable to suspect that in the latter case the magnet and the piece of steel are being *pushed* together in much the same manner as are the polished blocks?

Now, in the 20th century you and I *know* beyond much doubt what it is that is pushing the polished blocks together, whereas our forebears centuries ago, having no conception of the latent pressure in the all-pervading invisible gas we call the atmosphere, would have been totally mystified by the phenomenon. Difficult as it may be—and without wishing to offend—many of us are in exactly the same position today concerning the magnetic example.

These are the rather obvious questions I pondered on as a youngster, yet despite their elementary nature the answer to them is profound in the extreme.

As we have seen, the atmospheric pressure of a planet is caused by gravity and therefore it can be said it is a measure of the planet's material content, or massiveness. Thus it is that the atmospheric pressure at sea level on the earth may be only a fraction of that on a more dense planet.

To return to the outstretched hand experiment, we notice that we can feel the weight of our hand and arm and we have to exert muscular force to prevent it from *falling*. Now this may be quite obvious but what is not so is the fact that gravity is also trying to move your hand sideways but is prevented from doing so due to the fact it too exhibits an opposing force in the other direction which cancels the tendency out—in exactly the same manner as in the foregoing atmospheric pressure example—and likewise you are totally unaware of it. There are of course well-known classical experiments in which the *lateral* displacement—or *attraction*—between two massive objects can accurately be measured. Indeed, it is well known that an ordinary plumb bob weight will be displaced from the vertical if it is suspended at the side of a large hill or mountain, thus establishing beyond doubt the *omnidirectional nature of gravity*.

Occam's Razor

Although you don't have to fall off a chair to be convinced about the power of gravity, it is nonetheless a fact that compared with the electromagnetic and electrostatic forces, it is by far the weakest—as quantitatively indicated at the beginning of this chapter.

However, we are now venturing somewhat into the realms of uncertainty which have preoccupied and baffled particle physicists for years, in which there can be many false trails, so at this juncture we are well advised to take along what guiding tips are available. One of the most tried and best known to students is that of 'Occam's Razor.'

It is quite possible that both Faraday and Newton were influenced by the philosophical thinking of William of Occam, the fourteenth-century Franciscan, from whom the Occam's Razor rule was coined. He is quoted as stating that "Entities are not to be multiplied without necessity," or it could be restated; if two hypotheses—one more elaborate than the other—are offered concerning a certain problem or phenomenon, then Occam's Razor would require the less complicated hypothesis should first be considered. The relevance of this philosophical rule has been successfully demonstrated in countless applications ever since. It is a course the author took to intuitively, as many others have, and it is the directive I propose we should adopt from now on.

Magnetic and Electrostatic Phenomena

Sometimes it is good practice to go right back to square one and begin our thinking all over again. In the case of gravity we have no alternative, for as stated, very little is known about it; all we know is that matter seems to have the property of gathering together. For the want of better understanding, this has been described as 'attraction.' Einstein proposed that the phenomenon is due to a warp in space, however very few can understand this abstraction. To some this may sound impudent, but let's face it, it happens to be true. We know an ordinary bar magnet has this property to a marked degree over a piece of steel or iron. We note that in free space if the mass of the magnet and the mass of the iron are the same, the two would move together over equal distances and at uniform rates of acceleration. In other words, the motion is evenly shared. Now on the other hand, should we substitute for the iron another magnet of equal mass to the first and orient them with unlike poles facing, then they will also move together and the motion will again be evenly shared, but in this case, due to the greater *field strength,* there will be a correspondingly greater acceleration. All this is in accordance with Newton's laws of motion but note, the increase of *attraction* in the latter case has *nothing to do with the mass,* for there has been no change in this respect.

Similarly, an electrostatically charged body will exert an *attraction* on an uncharged body and if two such bodies of the same mass were in free space they also would move together at equal rates of motion exactly as the magnet and soft iron. Again, should the two bodies be charged with opposite signs, negative and positive, then the acceleration will be increased as with the two magnets of opposite polarity. Again there has been *no increase in mass*. Now by giving the magnets like polarity— for example north to north or the charged bodies a like sign, negative to negative— they will experience the same rate of acceleration as before, but this time *away* from one another. Currently, this is called 'repulsion,' but to this day science cannot explain it!

People the world over have spent their lives theorising, and whole volumes have been written on it, but let us have no doubt about it, we are standing on very firm ground when we say the world is still waiting for an explanation of this extraordinary phenomenon. Of course, there is the popular textbook theory in which each atom is said to behave as a bar magnet. In effect this postulates, if an electric current consists of billions of electrons flowing round a coil of wire produces a magnetic field within the coil—much the same as a bar magnet—then similarly an electron revolving in orbit around the nucleus of the atom will form a miniature magnet.

A bar of iron is said to be magnetised when the poles of the atoms are lined up one with the other. Apparently the only difference with other materials is, this phenomenon of lining up does not occur quite so readily as it does with iron or steel atoms. Therefore it is difficult, though not impossible, to align a magnetic field in, say, a piece of wood.

On the other hand, an electrified body exerts an attraction on a neutral body because it is said to have a surplus or a deficiency of electrons. The materials in a substance are said to be electrically neutral when the negative and positive charges within the atom exactly cancel each other out. In the simplest case, the hydrogen atom, the electron having a negative charge, cancels out the proton's positive charge and the atom is said to be electrically neutral, and the same is true of more complex atomic structures.

When electrons are removed from the atoms of a substance, the balanced state is upset and the material is said to be positively charged. Similarly, when the atoms of a body take on a surplus of electrons the body is said to be negatively charged. Textbooks inform the student such a charged body is *attracted* towards an uncharged body, or another body of positive sign, because the surplus electrons *want to dissipate themselves,* or the electron-deficient positively charged body *wants to borrow* some of the surplus electrons!

Again, within the structure of the atom itself, we find the same phenomena of attraction and repulsion, particle for particle, and wherever there is a moving electron there too is a magnetic field.

Summing up then, in all matter from the microcosm to the macrocosm the same basic pattern is there. Whether it be gravitational, electrostatic, or magnetic, there is this moving to and from tendency inherent in all matter, constantly obeying the inverse square law. But gravity is *always* there, be matter electrically or magnetically neutral or otherwise, it is always subservient to the force of gravity. Two magnets may repel each other, but the force of gravity is not one bit subdued by the magnetic effect. The same is true of two similarly charged bodies, gravitational attraction still remains operative between them.

The Electric Charge and the Electric Field

Around the early nineteenth century it was discovered that electric currents produced magnetic forces and that, conversely, when a magnet is moved about near a wire then a current is induced to flow. Textbooks also state that this interaction between electric and magnetic forces is central to an understanding of electromagnetic waves.

A convenient way of describing the forces which act between electric charges and currents is in terms of the field concept which holds that rather than say two charges attract or repel across *empty* space, it is claimed that every electric charge *produces* an *electric field* around itself, the strength of which diminishes with distance. The force that is exerted on a nearby charge is then attributed to the interaction between it and the field. Then we are told, this charge is also the source of its own field, which will react back on the former charge as well, the strength of the force being proportional to the strength of the field at that point.

Now, the student could not be criticised for asking, "If an electrically uncharged body does not radiate an electric field, but on being electrically charged it does, in what has the radiation been propagated?" Clearly, to be told that in some manner the field is a kind of *extension* of the charge itself isn't really convincing, whereas most lay people have little difficulty in relating to an all-pervading ether—in a 'first take' analysis anyway. I am tempted to go further and state, so long as physicists and engineers continue to accept as literal the electric charge/field concept, they will be severely handicapped when investigating new age energy and propulsion concepts. Meanwhile the world still awaits a unified theory.

A Magnetic Step Analogy Which Led to Wave Mechanics

It is anticipated that some may ask how it could be that a cosmological theory, no matter how valid it may be, could be developed by someone with an engineering background rather than that of a physicist. This is a fair question and as it may be of assistance to some aspiring young students I have considered it deserves some response here.

Due to the lack of fundamental training, for most of my early life—and to some extent even today—I have had to resort to approaching some questions concerning more obscure natural phenomena by viewing them in terms of observable, already well-understood facts, a typical example being that of magnetism. I reasoned that despite much mathematically supported theoretical treatment, no one was really telling me exactly what magnetic phenomena *really* are. I wanted to *know* why a bar magnet appeared to have two specific poles. Why these both attracted pieces of iron and other magnets with adjacent unlike poles, yet apparently repelled other magnets with adjacent like poles, etc. I reasoned that, after all, we appear to be witnessing a phenomenon which in its final analysis involved but two chief ingredients, i.e., space and pieces of iron.

Concerning this, I felt reasonably comfortable to view the situation in terms of simple well-understood hydrodynamic principles in which two plain cylinders representing two bar magnets, would be immersed in a medium—in this case a tank of water representing the surrounding space—with the hope that I might be able to repeat some of the observable magnetic effects and thereby gain a few analogous clues as to what was going on. In fact the idea does yield some fascinating similarities as follows. Fig. 3.3 (a) portrays a phenomenon called the Magnus effect (after its discoverer) which is well known to aerodynamicists. In this a cylinder, rotating about its long axis, is moved through the air as an airplane's wing in such a manner that the upper portion of the cylinder moves downstream and the underside advances. The resulting increase in velocity above the cylinder together with the accompanying pressure drop causes a very useful lifting component, L.

Figure 3.3 (b) shows the natural development of the idea in which a cylinder is caused to rotate without translational velocity. In this manner the air is dragged around due to viscosity and consequently flung out by centrifugal force, much the same as a catherine wheel, and it follows the same thing occurs if the cylinder is immersed in any fluid medium; in this instance I chose water.

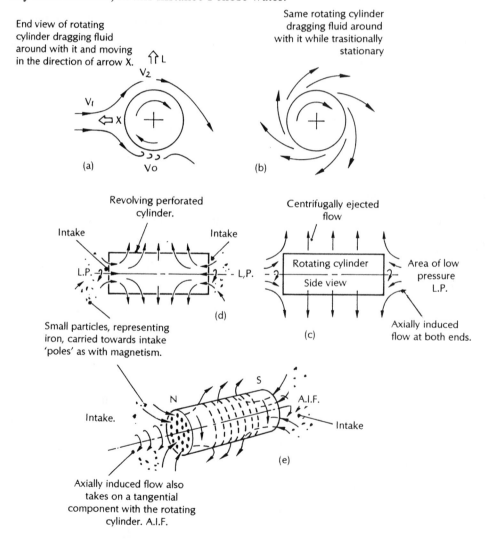

Fig 3.3. Arrangement and flow diagram of rotating perforated cylinder in a fluid

Fig.3.3 (c) represents a side view of such a rotating cylinder from which it will be apparent that if water is flung out along all, or part of its length, then that water has to be replaced due to surrounding water pressure and this replacement has to be at either end of the cylinder.

The next step in the analogy is to imagine the cylinder is hollow, open-ended and with perforated sides [Fig. 3.3 (d)]. It will be apparent that if the cylinder is rotated as before, nearly identical effects take place, but now the water enters both the open 'intakes' and is centrifugally discharged along its length, as in (b) and (c), Fig. 3.3 (e) being a more graphical representation of the general idea. In effect we have created a simple double-sided centrifugal pump with radial discharge and it will be understood that there will be an accompanying local pressure drop at the intake mouths and further, any nearby particles suspended in the fluid will be sucked into the intakes and arrested by a filter gauze. For convenience, the ends of the cylinder are labelled North and South, as with magnets and we can now examine some of the attendant effects which are quite fascinating, and as will be seen later, not without important significance.

Fig. 4.3 (a) represents two such rotating cylinders placed end to end and a short distance apart, the direction of rotation being the same for both. In this case due to the proximity of the *inner* intakes, there is an additional pressure drop, thereby effectively creating a pressure differential which causes the freely suspended cylinders to move together, to form one elongated double-sided centrifugal pump. In exactly the same manner as observed in magnetic 'attraction,' in which two unlike pole magnets join together to become one magnet.

If the right-hand cylinder is kept rotating in the original direction as in (b), but is now longitudinally rotated so that the two 'south' facing ends are in close proximity, it will be apparent that the cylinders are now contra-rotating. Moreover due to the fact that the fluid at the intakes takes on the rotary swirl of the cylinders, an antagonistic condition is created with consequent loss of velocity and accompanying pressure *rise.* This increase in pressure now represents a pressure differential as before only now in the opposite sense and the two cylinders move apart, just as in magnetic 'repulsion.'

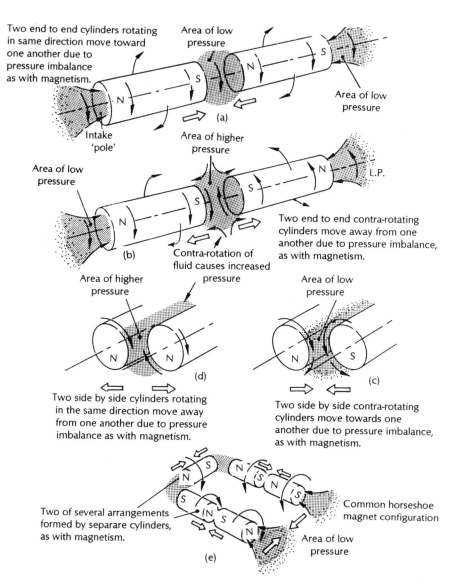

Two end to end cylinders rotating in same direction move toward one another due to pressure imbalance as with magnetism.

Area of low pressure

S

N

Area of low pressure

Intake 'pole'

Area of higher pressure

(a)

Area of low pressure

S

N

N

L.P.

Two end to end contra-rotating cylinders move away from one another due to pressure imbalance, as with magnetism.

(b)

Contra-rotation of fluid causes increased pressure

Area of higher pressure

N   N

(d)

Two side by side cylinders rotating in the same direction move away from one another due to pressure imbalance as with magnetism.

Area of low pressure

N   S

(c)

Two side by side contra-rotating cylinders move towards one another due to pressure imbalance, as with magnetism.

Two of several arrangements formed by separare cylinders, as with magnetism.

N   S
S   N
N   S
N   S

Common horseshoe magnet configuration

Area of low pressure

(e)

Fig 4.3. Main magnetic effects simulated by rotating perforated cylinders in a fluid, in this case water.

We can now complete the experiment by noting that if two cylinders are longitudinally aligned side by side at a short distance from one another with north and south ends adjacent as in Fig.4.3 (c), again due to a concordant sense of rotation in the fluid, there is an increase in velocity set up between them together with an accompany pressure drop, resulting in a movement towards each other as with magnetism. Should we now longitudinally rotate one of the cylinders to give north/north, south/south ends proximity as in (d) there again exists a reverse flow condition with resulting decrease in velocity and increase in pressure, which forces the cylinders apart as with magnetic repulsion. Note, during all the above manoeuvres the small suspended particles will continue to be 'attracted' towards the poles.

It should be noted that, as with two bar magnets, the foregoing effects are brought about *merely by polar orientation*. Also, if at first acquaintance the experiment may appear to be peripherally at variance with observable magnetic effects, closer examination in fact reveals a surprisingly close agreement.

Fig. 4.3 (e) illustrates that even the observed characteristics of the humble 'horseshoe' magnet are adhered to, thus supporting the author's belief that there is a kind of helpful duplicity in natural phenomena which can be—and has been to me—of inestimable assistance to the enquiring mind. It is hoped this original example will prove to be no exception in helping the reader to relate to the developments in the following chapters, whilst keeping in mind the fact that at this stage we are still dealing with analogies.

Now here we are at square one; shall we for the moment then leave the well-trodden paths of established scientific pronouncement and give credence to Occam's Razor rule of economy? By applying this rule we would in the first instance be prompted to suppose that magnetic, electrostatic and gravitational phenomena might indeed have one common cause. In which case we would still be left with another two alternatives—one of which is claimed superfluous by Occam's Razor—that of 'attraction' or 'repulsion.' Later on I propose to show this is exactly so and is due solely to orientation between bodies of matter, in similar vein to the foregoing hydraulic analogy. Naturally we would have to talk in terms of 'attraction' and 'repulsion' in order to designate the relative motion, but due to the economy rule, our hypothesis has been drastically simplified. We are left with *matter* and relative *motion*. And if Occam's Razor is to be faithfully adhered to, we should consider the hypothesis that magnetic and electrostatic attraction and repulsion effects are a kind of selective gravitation. In modern relativistic thinking there is no room for the outdated conception of an all-pervading ether and of course science is carrying on quite nicely without it, largely by the substitution of complex alternative abstractions. But if Occam's

Razor requires simplicity, then I suggest for a while we should keep our ether, for we shall see that it becomes the most important issue of them all.

Matter

Insofar as the phenomena of magnetism, the electric charge, and gravity are inherent functions of matter, any meaningful analysis of one aspect inevitably leads to the other. What really is this stuff which we are apt to take so very much for granted? We are all familiar with the basic premise that the atom is the building block of all solids, liquids and gases and each atom is said to be composed of a nucleus around which electrons revolve. The nucleus itself is made up of protons, positrons and neutrons, etc., which vary in number depending on the type of material involved. It is also accepted that these subatomic particles also carry electric charges and that they have microscopic dimensions compared to the size of the atom itself. As stated, the accepted model of the hydrogen atom basically consists of one proton in the nucleus which is positively charged and one orbiting electron which is negatively charged. It has been said that if we imagine the nucleus to be situated at the centre of London's Albert Hall, then the orbiting electron—travelling at near the speed of light and invisible at this scale—would be situated at the outer walls! Both these particles are said to represent in some respects packages of electric charge, while the electron itself possesses only inertially acquired mass, which implies it should vanish when it is stopped! In a word, the hydrogen atom and for that matter all atoms, are nothing but 99% space and what stuff resides therein is more akin to pure energy and nothing else. Thus the page upon which I am writing is in reality an almost empty void despite its apparent solidity.

The Permeability of Matter

Sometimes at lectures, in an attempt to visually convey a measure of the sheer tenuity of matter, I have resorted to a very basic though limited atomic analogy of mine, repeated here for those who would wish it. In this we have to imagine the hydrogen atom to be represented by a very small weight attached to a length of string several feet long being whirled round by the hand; the weight representing the electron and the restraining hand represents the nucleus. In this 'thought' experiment we have to imagine the restraining hand and the string to be replaced by a force akin to magnetism.

In the mind's eye we can now imagine that by computerised control we could have several such weights synchronously whirling in different planes without colliding. Should the orbital speed be sufficient, the light reflected by the weights would give an impression of interlocking rings. Furthermore, given that the number of weights could be further increased, we would get a visual impression of a solid sphere, if we could prod the thing we would feel and hear it; in a word we couldn't be blamed for mistaking this allusion for a globe or sphere which in reality is composed of only a few small rotating weights. However, that's the easy bit. In order to complete this simple analogy requires a little more mental gymnastics. In this we create two such illusory spheres and require even more ability from the computerised controllers, in that just as the orbits of the weights in the individual spheres were synchronised to avoid collision, so the orbits in *both* spheres are synchronised, thus enabling the spheres to merge one with the other. Thus, *theoretically* speaking, in this analogy there is ample space for two or more separate systems in which to all appearance seems to be one solid body. There will be a further application for this analogy in Chapter 5.

It must be reemphasised that the whole point of this analogy is to illustrate the extreme permeability of matter, even in this instance of hypothetical billiard ball mechanics.

In 1900, Max Planck's Quantum Theory negated the idea of orbiting electrons in distinct locations around the nucleus. As if this wasn't enough, it was then established that even the protons and neutrons forming the nucleus were themselves composed of even smaller particles known as Quarks. As mentioned earlier the situation was further complicated when it was discovered that in some circumstances the electron displayed both particle *and* wave characteristics, now identified as wave/particle duality.

In order to explain even more obtuse phenomena, modern physicists have invented an activating process tentatively designated the 'messenger.' In the case of gravitation, for instance, it is the hypothetical 'graviton.' For the electrostatic fields the 'messenger' is the 'photon,' both of these are claimed to travel at the speed of light, both having a rest mass of zero. In more recent times there have been various 'stabs in the dark' attempts to lend some kind of basic structure to the Universe, so perhaps it is inevitable that attention will be drawn to any correspondence between these and that offered in the next chapter, no doubt the currently popular 'Cosmic Strings' idea being a more obvious example. However, logically, if a choice has to be made between a coherent versus random theory, then surely we should bear in mind there is nothing random about the precise nature of the velocity of light!

However, before concluding this section, there is one other most important aspect of matter which should be briefly included, and that is energy.

Matter and Energy

Now in order to condition ourselves for the conclusions in this book, it may be as well to consider some facts about matter and energy in terms of power. There are three main levels known to man at which energy can be extracted from matter. Two of them are currently fairly well understood, the third is not. To deal with these in order, we have first the 'thermal' process in which radiation is emitted by the chemical change known as burning, over the parts of the electromagnetic spectrum which give us light and heat.

The second level is the 'thermonuclear' process, in which yet more heat is liberated by nuclear reaction; the level at which even the waste products left over from the thermal chemical process are still capable of yielding an enormous amount of comparable energy. In other words, the waste products given up by that screaming jet fighter as it streaks overhead, and the latent energy in the exhaust residue of a rocket blasting spacewards are still capable of yielding enormous quantities of energy, nuclear energy, which if unleashed would fly many more fighters and send many more rockets to the moon and back many times over.

The third level is the *direct* conversion of mass into energy, a technological wonder of the future; this is the domain in which some physicists are contemplating the final level of exchange on the *physical* plane. [Fig. 5.3 (a)] It can be shown that the fusion of only 1 gram of hydrogen into helium will yield *200,000 kilowatt hours* in the form of heat and radiation. This liberated energy being the result of the 'mass defect' or the minute difference when the helium nucleus is formed. It can also be theoretically shown that when 1 gram of ordinary mass is *completely* converted into energy, no less than *25 million kilowatts* become available for one hour! The relationship of these three levels of exchange is shown more graphically in (b).

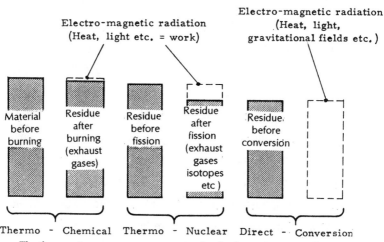

The three main mass-energy conversion levels shown diagramatically
as it would be impossible to show the true relationship in this scale.

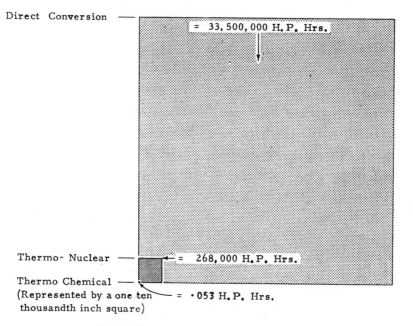

Fig 5.3. Theoretical energy relationship at the three
main levels of exchange, for an expenditure of 1 gram.

The link between space and matter has long taxed the imagination of many original thinkers (in addition to those highlighted in the Appendix), these include Nikola Tesla, Alexander Schnetski, Hal Puthoff and Jim Briggs, etc. They all believed that space itself is a sea of inexhaustible energy. In order to find some coherence between these ideas and known established scientific fact, reference has been made to 'energy locked in a vacuum.' More recently, some physicists have invented the term 'zero point energy'; the situation which calculably exists when the temperature of matter is reduced to zero. Apparently it is visualised as a domain which is distinct from ordinary matter. In the following chapter reasons are offered to indicate that zero point energy may be merely another level in a totally encompassing energy field. Not least, it will be shown that the long sought-after 'Rosetta Stone'—proffered by some scientists who are desperately seeking the unification of the existing divergent laws—is probably much closer to hand than they are apt to presently believe. Despite suspicions to the contrary, those who *have* been open-minded enough to take the trouble to investigate the claims of the aforementioned pioneers, together with others, have admitted with bewilderment, these claimants all had one thing in common, they *did* produce an unaccountable amount of surplus energy . . . free, clean energy!

Unfortunately for particle physicists, all these developments represent the mere tip of the quantum mechanics iceberg currently being contemplated and with that in mind perhaps we might share an affinity with Dr. Leon Lederman when (as quoted in Chapter 1) he said, "We have reached a crisis in physics where we are drowning in theoretical possibilities not based on a single solitary fact." This sense of acute exasperation might in no better way be epitomised than by Sir Fred Hoyle, who when referring to quantum mechanics in his fascinating book *The Intelligent Universe* said he could well remember when one of its strangest implications hit him as he sat on the banks of the River Cam just outside Cambridge in 1938. He had only recently won a sought-after research prize in this subject and was about to become a research student of the great Paul Dirac (one of the foremost physicists of the twentieth century). He had asked himself what terrors quantum mechanics could hold for him. He had to admit to plenty, and as he sat waiting for afternoon tea at the local inn, he suddenly saw that up to this very moment he hadn't understood it at all! He said that he realised that unknown to him then he must have felt exactly like Erwin Schrodinger—who, he had no doubt, was similarly appalled by the magnitude of the problem—had exclaimed, "I don't like it, and I'm sorry to think that I ever had anything to do with it."

For my part, I shall always count myself fortunate when in 1953 an unknown author presented me with the Unity of Creation Theory, which I had no difficulty in likening to a multi-dimensional canvas on which I could paint a far more extensive range of colours in the form of natural phenomena. Granted, here and there my choice, or relative positioning of these 'colours' may not be quite as they should, yet somehow I feel content that for a relatively junior student the overall picture I have come up with is not too bad at that. For the benefit of those who are unacquainted with it, the Theory of Unity has been included in the following chapter just as I received it in 1953.

[1] This is noticeably more prevalent in the USA.

[2] The author's son Gary quite independently arrived at exactly the same conclusion long before me.

[3] Author of *The Birth of the Quantum*

4

TOWARDS THE UNIFICATION OF GRAVITY AND PRIMARY ENERGY

At this time I cannot be certain of the veracity of my own interpretations, but it seems to me that for anyone to obtain positive results, such as those mentioned in the previous chapter, with experimental apparatus based on little more than a "stab in the dark" hypothesis of 'space energy,' suggests that either it is a comparatively easy thing to achieve or it borders on the miraculous. But as we shall see later, this sort of thing really does happen! Though logically we might expect that definitive results would be more likely if such empirical work was more conceptually based. Therefore, in this chapter we shall be attempting to establish this by adopting something approaching a theoretical metamorphosis in which particle mechanics transcends towards pure wave mechanics. So for lay readers, a few introductory words may not be amiss.

Throughout the previous chapter we saw how it is more convenient to describe gravitation in terms of an ether than not to accept ether at all. As a first-take analysis, in purely mechanistic terms, this inherently suggests a kind of 'pressure' from without rather than a 'pull' from within—as illustrated by the two ground cubes of metal analogy. We saw how the cubes were held together because the outside pressure of air was not contested by an equal pressure between them. Textbooks would describe this as a 'suction' effect, but this term is purely a descriptive convenience, for there is no such physical thing. In a word, so-called 'suction' alone cannot exist without the manifestation of *pressure*. If by withdrawing a piston in a pump we experience a 'suction' effect it is due solely to the fact that the existing ambient pressure is trying to *push the piston back in*; in other words we have created a pressure differential. Thus it is easier to imagine an ether pressure pushing matter together, but we cannot even begin to imagine an ether pull from within—as in the sense of elastic bands. [Fig. 1.4]

To consider a boundless space bespangled with matter in varying degrees of formation upon which is exerted a pressure from without leads to the obvious question: What is the origin of this pressure and how is the layman to visualise it free from mathematical entanglement?

The following theory offers a possible explanation, the nature of which conforms to a general pattern displayed by nature, that is, circular motion. We have seen how some of the greatest scholars have postulated a kind of unity in the universe, from Faraday to Einstein and many others. In 1953, in one of the most fas-

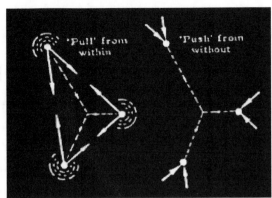

Fig 1.4. Is gravity a 'push' from without or a 'pull' from within? This sketch represents a three-body situation.

cinating and complete works by an Englishman, Antony Avenel of Yorkshire presented an article titled "The Unity of Creation Theory."[1] The following is a reintroduction to it.

The Unity of Creation Theory

The recent correspondence in technical magazines seems to show that many readers feel the need for something less coldly mathematical than Einstein's 'Theory of Relativity' and his subsequent theories. Few suggest that Einstein's brilliant calculations and theories are faulty, yet by themselves those essays in pure logic are not comprehensible to the average person.

One cannot gain the *mental picture* of Einstein's theory because the theory is not in a form which leads to a mental picture. If you read the test performance figures for a new aeroplane, you will know a lot about what the plane will do, but you will be unable to visualise whether its lines are beautiful or ungainly, or anything of its appearance.

Einstein's theory was before its time. The calculations of, for instance, the precise amount that a material contracts along the direction of its movement seem to be out of place before it has been explained why the contraction occurs.

I suggest that there has been too much mathematical jigsaw puzzle making and solving, and that formulae have been put forward which, though probably correct, are by themselves without very much meaning to the intellect.

The Michelson-Morley Experiment

This experiment aimed at finding the speed of the earth through the ether. Scientists had assigned to the ether descriptions ranging from an elastic solid to a rarefied gas. If our speed through the ether could have been determined, it would help us to understand—among other things—whether we are labouring through a mass like black treacle, or wafting our way through a substance as thin and delicate as perfume.

Those who rely on this experiment, or on similar experiments make an assumption which I believe to be false, that is, that the ether is a three-dimensional substance—such as gas. Only if the ether was a material substance would the passage of the earth through it cause an ether drag, or an ether wind, which could be measured.

The result of the Michelson-Morley experiment showed—apparently—that either there was no ether, or if there was an ether the earth was not moving through it. Neither of these conclusions seemed to be probable; it would be unlikely that the earth should remain stationary in space when all other observed heavenly bodies were moving. Nor was it likely that there was no ether, for how else could the passage of rays through space be explained?

As neither of these conclusions could be welcomed, it was later suggested that the Michelson-Morley experiment really did show a positive result, but that the measuring rod in the direction of the earth's movement through space contracted by an amount exactly sufficient to remove the positive result from being apparent (the Lorentz Contraction). The proposition was of course that all materials contracted in the direction of travel; the supposed contraction was not confined to the measuring rod in the Michelson-Morley experiment.

The Lorentz Contraction at first sight seems to be an artificial and farfetched theory, yet I think that those who have studied the calculations, and those who care to do so, will agree that the contraction must be accepted as something which actually does take place.

The Theory of Unity suggests reasons why the Lorentz Contraction takes place.

Theory of Unity of Creation

The interest shown lately in the physical world prompts me to offer an outline of that part of the theory which affects this subject. The theory suggests, among other things, why the phenomena forecast by Einstein's theories take place. It is unsatisfying to be told that time slows when you travel through space, and even to be informed of the precise amount by which it slows compared with your velocity, before any attempt is made to explain what time is and why it is capable of slowing.

The following statements and arguments are set out in rather a dogmatic and

The Cosmic Matrix According to the Unity of Creation Theory

Hypothetical portrayal of the globular formation of the Cosmic Matrix depicting only a few space forming rays in one plane only. A point of space is created at each intersection which carried to its logical conclusion produces a homogenous 'ether' with an infinite weave or structure and probably a diameter of over 16 billion light years.

The formation of space/time continuum according to Antony Avenel oversimplified form, which I hope will be excused, in order to try to offer an outline of the theory which can be followed without undue effort.

The theory anticipates the ultimate result of the fact that research discovers one unity after another in physical phenomena. One is led to expect that before long it will be proved that there is one basic building material for the whole universe. I do not pretend that there is sufficient data available at present to prove the theory fully, but there are many indications that it is an anticipation of what will be proved by, let us say, the year A.D. 2000.

The theory that I put forward is that the ether and space are the same, and that space is formed out of nothing by a grid of extremely high frequency rays (probably having a wavelength of less than 10^{-13} cm). Space must be distinguished from 'nothing.' Space—even if it is empty—possesses the qualities of length, breadth, thickness and time. 'Nothing' has no qualities whatsoever, and cannot support any material or ray. In other words, creation of the universe takes the form of making space out of 'nothing,' and the method adopted for making space is a network or grid of rays, which I call 'creative' rays.

Outside the Universe

Taking 'the universe' to mean all created space, there is 'nothing' outside the boundaries of the universe. The old problem of imagining the boundaries of the universe, outside which stretched empty space—which space must have boundaries, and what was outside that?—should not arise. 'Endless space' is a contradiction in terms. Space has dimensions and boundaries and cannot be endless. The hand of creation has not touched the 'nothing' outside the boundaries of the universe, and that 'nothing' has no dimensions and therefore no boundaries.

To put it another way, space is positive creation, while 'nothing' is the absence of space, and thus purely negative.

You cannot visualise 'nothing' for obvious reasons; it has to be accepted. If anyone particularly wishes to try to relate it to human experience, it could be said that they have had more of it than they have had of space and time. It is what they experienced or did not experience, before they were born.

Space or ether is formed by the creative rays which emanate from one source in all directions and in all planes [Fig. 2.4 (a)]. Each creative ray covers a circuit from source back to source, and each circuit is probably the same size. In this way space with boundaries of globular shape is built, and whatever point is taken in space, creative rays travel in all directions towards the source, as in (b) and more dimensionally in (c).

By the word 'source' I do not imply that the formation of the creative rays operates in only one direction in each circuit; the action may be alternating.

Light is a Modulation of the Creative Rays

All rays of whatever frequency, visible or invisible, detectable or undetectable, are modulations of the creative rays, in the same way as a high-frequency radio wave is modulated by a musical note. As a radio carrier wave can be modulated by a number of separate notes, so can the ether carry between the same two points any number of waves of differing frequencies.

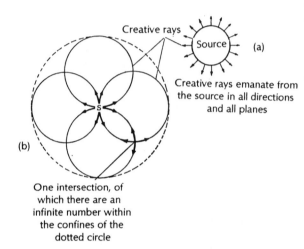

It would appear that rays or modulations are always caused by a disturbance in three-dimensional material, and that they are only of consequence when they encounter other such material. When a ray travels through space it is merely a slight modulation or disturbance of the creative rays and of no importance.

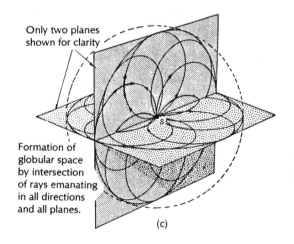

Material Objects

The atom is the building material for all solids, liquids and gases, and each atom is composed of a nucleus round

Fig 2.4. The formation of Globular Space according to the Unity of Creation Theory.

which revolve electrons at distances from the nucleus which vary with the type of atom. I submit that the atom is not solid fundamentally, but that it is composed of modulations of the creative rays in three planes. Although a modulation is normally a ray which travels in all directions from *its* source towards the source of the creative rays, the chord of modulations form-

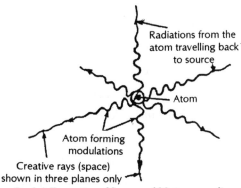

Fig 3.4. Formation of Space and Matter according to the Unity of Creation Theory.

ing an atom are locked together in three planes. This lock prevents the modulations travelling in opposing directions as rays. Does not the release of atomic energy show the very close relationship between atoms and rays? [Fig. 3.4]

The main point which I want to make is that rays and atoms are both modulations of the creative rays, the former being simple modulations, and the latter being complex and static ones.

An atom could in some ways be compared with a ripple caused by a stick in a smoothly flowing stream of water. It remains the same in appearance, yet it is formed from a constantly changing medium. If this is correct, the universe is made from the same medium throughout, and what appears to be empty space between the Earth and Mars is in reality a connecting medium.

Time

I suggest that time is the effect on our minds of the frequency of the creative rays. If the atoms out of which our brains and bodies are made are formed out of the creative rays, we cannot but be aware of the alternation of the creative rays. We cannot escape from time unless we also escape from space, or, in other words, cease to exist.

It is impossible to look either backward or forward in time from a fixed position in space. If we could travel at the speed of light and thus 'keep up with time' we should probably cease to be three-dimensional, which would not assist our observations! In any case, we should find ourselves in a different position in space, so we cannot by any means foresee what is going to happen, or look back on what has happened on earth.

Alternation of Time

If we were to travel at a very high speed—a substantial proportion of the speed of light—the frequency of the creative rays in the direction of our travel would be increased, because we would be travelling relatively to the pulses of the creative rays. It can be envisaged that something akin to the Doppler effect would take place, with the result that our basic time would be increased in frequency. We should not be aware of this, because the frequency of the creative rays is our only standard of time, and there is nothing *nearby* against which we can test this standard. But a stationary observer could, by rays of light, calculate the difference between our time and his time; he would say that our clock was going slow compared with his clock, or that our basic time frequency was quicker than his.

Clock time is our way of counting the number of pulses of basic time. If basic time frequency increases, clock time still counts as one million pulses what are now, say, two million, and the clock time appears to be going half speed.

Some space travel enthusiasts consider that if you could travel fast enough in a spaceship, you could spend twenty earth years away from our planet and come back only a year or two older than when you left. If this is calculated using basic time-space, it is found that the effect on your body and the impression on your mind, is exactly twenty years' worth of earth time, and that you could not therefore enjoy almost perpetual youth by this very inconvenient method.

Contraction of Length

If an atom moves along the creative rays, the increased frequency referred to before results in a shorter effective wavelength of the creative rays, which decreases the measurement of the atom in the direction of its travel.

For the purpose of simplicity, take it that the material length of an object is formed by the wavelength of the creative rays, while basic time is the frequency of the creative rays; then wavelength times the frequency of the creative rays will remain constant at whatever speed the object travels, because as the frequency increases the wavelength decreases. The product of the length of the object and basic time is unaffected by the velocity of the object and it is this product which gives to our minds the impression of time and of the proportion of the object.

The creative rays present existence or the possibility of existence, and time and space are a division of that presentation. In whatever proportions the division is made, the whole remains unchanged.

Rays and Materials are Temporary

I would now like to meet the objection of those who say that it is just as diffi-cult to believe that the creative rays travel through 'nothing' as it is to accept that light travels through space without an ether to carry it. My reply is that the theory holds, that the creative rays create space not casually but permanently; their cause is not casual, like the cause of a ray. The theory proposes that rays and materials are casual and temporary modulations or disturbances of the creative rays. It would seem unrea-sonable to believe that a special act of creation is necessary every time you choose to switch on an electric torch. The theory of unity holds that you, by switching on the torch, are able slightly to modulate the creative rays, which are permanently present, and that the casual phenomenon of visible light is the result.

Gravity

It is usually accepted—to put it basically—that if in space two masses exist, they attract one another. I suggest that this idea is wrong, and that it is impossible for a material object to emit rays which pull another object. Nor is there anything other than a ray which could exert the supposed pull. Rays can exert a small amount of pressure on an object in the direction of the ray's travel, but they cannot pull.

An alternative theory is that gravity is due to an increasing velocity, and the anal-ogy of a lift rising at constantly increasing speed is often used. If a person in this lift released a pencil, it would appear to that person to fall to the floor of the lift and he might well consider that the pencil was attracted by the floor. If this is the explanation, why does gravity act in more than one direction? It requires adjustments which seem to me to be very artificial to answer this.

The theory of unity explains gravity as the material version of the natural travel of a ray towards the source of the cre-ative rays. The modulations forming an atom tend strongly to split up, to break their three-dimensional bond and to travel in all directions, like ordinary rays towards S.

Referring to Fig. 4.4 (a), S represents the source of the creative

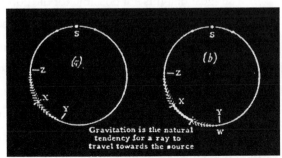

Gravitation is the natural tendency for a ray to travel towards the source

Fig 4.4. The function of Gravitation according to the Unity of Creation Theory.

rays, and the circle represents the circuit of one creative ray. While X represents the place at which an atom is formed by the intermodulation of the creative ray shown on the diagram with the creative rays in other planes; the latter cannot clearly be represented on paper, nor of course can an attempt be made to draw to scale.

The tendency of X to act as a ray and to travel to S via Y and Z, and via other planes, is nullified by the three-dimensional strength of X. The modulations of the creative ray start for practical purposes at Y and Z, but they do not interlock with modulations in other planes until X is reached. These preliminary modulations in ray form I will call extension modulations; some of them are of measurable frequency, others are of a frequency too high to be measured by a material device.

So long as X is undisturbed, it remains still in space, the tendency to travel to S via Y being balanced exactly by its tendency to travel to S via Z. Its tendency to travel to S in other planes is also balanced. But, referring to Fig. 4.4 (b), if in the position of W—before the extension modulations of X have for practical purposes faded—another atom is formed, the extension modulation of X is interfered with and unbalanced between X and W. The extension modulation of X in the direction of Z is unchanged; XZ and XY are now no longer balanced, and as a result X moves towards W; W also moves towards X according to the laws formulated by Newton, or approximately so.

Although X moves towards W, it is not *attracted* by W, any more than light from the sun is attracted by the earth. (Here I am ignoring the almost negligible element of gravitation between a material object and a ray: the reason why light travels from the sun towards the earth is not because of mutual attraction between light and the earth). X moves towards S via W.

Magnetism

I suggest that it is not possible for the north pole of a magnet to emit rays which attract the south pole of another magnet, and repel the north pole of another magnet.

The travel of one magnet is not towards another magnet but towards S. Some atoms of iron are arranged, or can be arranged, so that the extension modulations are not the same in all planes. This lack of symmetry can be encouraged by electrical means. It is quite possible that a single magnet removed from a powerful gravitational field would move through space of its own accord. A single magnet on or near the earth is prevented from moving by the gravitational field of the earth—that is, it is prevented from moving through space of its own accord. If another magnet of opposite polarity or a piece of iron is placed near the first magnet there is apparent attraction, but what actually happens is akin to gravitation. The first magnet moves towards S until it reaches the second magnet or piece of iron. The strength of the magnet probably depends on the number of atoms in the magnet which have unbalanced extension modulations, the degree of lack of symmetry in each atom remaining constant.

Electricity

I suggest that this is a general disturbance of the extension modulations.

Flying Saucers

True 'flying saucers'—that is, those which are not the results of the imagination of the observer—are vehicles which are based on the principle of unbalancing the extension modulations of material carried in the vehicles.

Reality

The questions arises: "Are these changes in time and space real, or are they only *deemed* to happen?"

The answer to this is, what you and I and everyone else are concerned with is basic time times length, representing the whole effect of both the frequency and the wavelength of the creative rays. In judging reality before our eyes, we are not concerned with the division of time times space into time and space.

If you want to listen to a concert on the radio it makes no difference to you whether the programme is carried to you by a 500-metre carrier wave or a 1,000-metre carrier wave, and you could detect no difference in the reality of the reception. You might then say that there was no real difference; but an engineer who is more interested in the method of your hearing the programme than in the programme itself would say that one programme was the result of modulating a carrier wave of 500 metres wavelength and frequency of 600kc/s, while the other programme was brought on a carrier wave of twice the wavelength and half the frequency. To the listener who was unable to go further into the problem than to hear what came out of his loudspeaker, the programme would be the same.

The answer then is, shortly, that although the change does actually take place in time and space, it is not real in the sense that it could be observed by a human being living within the sphere of the change, for such a person has not the means to measure basic time or basic length as an engineer can measure the wavelength and the frequency of a radio carrier wave.

It is now nearly four decades since the publication of the "Unity of Creation Theory" in which Avenel stated, "there are many indications that it is in anticipation of what will be proved by, let us say, *the year A.D. 2000.*" During the intervening time the present author has indeed witnessed many more of these indications. I have met many people who continue to be moved and impressed by the sheer elegance of this remarkable completely encompassing theory. Also it must be stressed that the following observations and projections are my own, with which Avenel may not necessarily have been in agreement. Therefore I take full responsibility for any inaccurate interpretation on my part, although I like to feel that to a useful degree I have contributed a little towards the *"proof by the year A.D. 2000."*

[1] Occasionally referred to as "The Theory of Unity"

5

THE COSMIC MATRIX
Was There Really a Big Bang?

In the intervening forty years since I received the Unity of Creation Theory, and not having been in touch with its author, I suppose it was inevitable that without deliberate intent, I should constantly find myself expanding on the general theme.

One of the more obvious of my additions was the abridged title *Cosmic Matrix*, for it seemed to me that was exactly what Avenel was describing. The remainder of this chapter is devoted to some of my inclusive additions.

The Inverse Square Law

The Theory of Unity is inherently consistent with the Inverse Square Law and in order to proceed further, one of the simplest ways to imagine the inverse square effect is the optical application, in which a lamp throws an image on to a screen as in Fig. 1.5 (a).

First consider the larger screen designated Area 1, the dotted lines portraying the rays of light from the lamp, which in this instance is placed, say, two feet from Area 1.

Now should this distance be exactly halved by moving the screen nearer to the light source, then obviously the rays of light intercept the screen at different positions producing the Area 2, which due to the inverse effect happens to be exactly one quarter of Area 1, i.e., one square foot. As an ex-

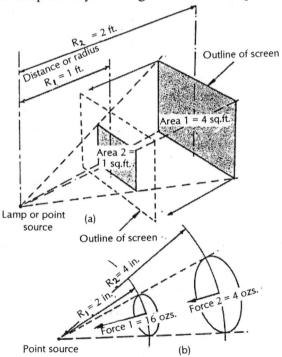

Fig 1.5. The inverse square law explained in terms of light.

ample the units of *area* and distance were chosen, but the same rule holds for *force* and distance.

Fig. 1.5 (b) also represents a radiating point source, but this time it is radiating not light but a field of *force*, it could quite easily be a magnetic, electrostatic or a gravitational field. Given the force at one distance and by applying the same rule, the force at any other required distance can be found.[1]

Therefore the rule of the inverse square law: double the distance and we get one quarter of the force, or halve the distance and the force is quadrupled. Here we have halved the distance for convenience; the same will hold true for any other values we care to employ.

Fig. 2.5 shows how even the inverse square law phenomenon can be graphically explained by the Theory of Unity, in that the closer two bodies approach each other, so the more 'C' rays there are unbalanced, and therefore the greater the force between them.

Inertia

The theory can now be expanded further by stating if gravity is one of the most enigmatic forces in nature, surely the *property* of inertia runs a close second. Indeed,

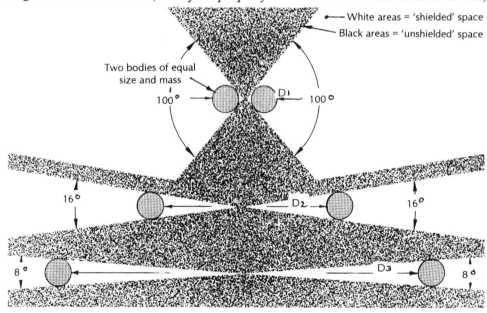

Fig 2.5. Shows two bodies of equal size and mass situated at three different distances from one another. The white triangles represent the consequent increase in effective shielding and therefore the increase in 'attraction' as the distances get less according to the inverse square law.

Einstein was one of the first to propose the two are inseparable. So to suggest that the existence of inertia—in the form adopted in classical physics—is wrong and propose there is something akin to a 'higher octave' of matter in space (Avenel's Creative Rays) might be considered contentious, but it should not be overlooked that in the beginning, the term inertia was invented for a condition we don't really under‑stand.

What is inertia? Can we handle it or buy it by the pound? Has anyone seen it? The textbook tells us it is the property which a body 'possesses' by which it

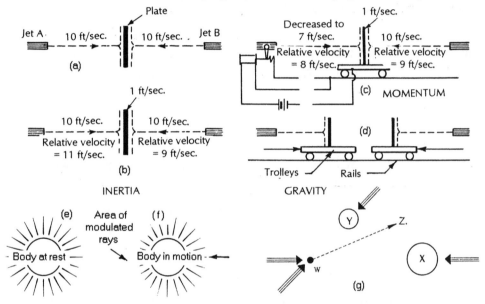

Fig 3.5. Motion, Inertia and Gravitation are manifestations of modulated space.

shows a 'reluctance' to be moved from a condition in rest, or 'reluctance' to be brought to rest when 'possessing' 'momentum.'

Tell some physicists that there may be a state, or a 'higher octave,' of matter and we will be received with raised eyebrows, yet no one shows any alarm whatsoever when we say a body 'possesses' such an intangible thing as inertia! In actual fact the one is just as intangible as the other. Let us reflect; in the first place a body is said to be at rest when it has no relative motion to another body, but is there such a state? Not that we know of, indeed it is most unlikely that there is a single body in the whole of the known cosmos that has no relative motion of some kind.

Consider a body in space, there being no planets, suns, nothing save the one body. It is obvious such a body can be moving only in respect to something else, but there is nothing else, so the body isn't moving! According to classical physics the body still possesses inertia. It will *stay where it is* unless a force is brought to bear on it!

Now the body is surrounded by nothing—we will accept for the time being the modern trend of thought that there is no such thing as an ether—how then can we visualise the concept of a body possessing of itself a reluctance to move *unless it is fixed*, or in some way related to something else? Surely we must come to realise that the surrounding space *is* that something else? Then it follows that if the body of matter—the term given by us to a collection of non-material *forces* called atoms—is affected by space, then in some way that space must be composed of a 'higher octave' of matter.

Let us examine another simple analogy, for in it we may find several interesting possibilities as far as inertia is concerned. Fig. 3.5 (a) shows a plate mounted upon a trolley on which impinges two opposite and equal flowing jets of water, thereby holding the plate in a state of equilibrium. It is quite clear that the force acting on the plate is only proportional to the mass flow of water issuing from the jets and it follows, any increase or decrease in the jet velocity will result in a greater or lesser force.

Should we try to move the plate in the direction of one of the jets as in (b), it follows there will be a relative increase in velocity on that side and a corresponding relative decrease in velocity on the other side. Therefore, the plate will experience an unbalanced pressure and will show a *reluctance* to move, representing inertia.

But a body in space will only show a reluctance to move initially, thenceforth it will continue to move. Therefore, we must modify the analogy to suit this requirement; easily achieved by making jet A in (c) controllable by means of a solenoid-operated cock, and the solenoid in turn controlled by an electrical contact formed between the wheels of the trolley and the rails.

Now, when we try to initiate a movement, there will still be the reluctance to move as in (b), but immediately after this stage has been reached, the wheels complete the electrical circuit, the velocity of jet A is decreased and as will be seen, the relative velocity of it will still be less than the relative velocity of jet B. The plate, after exhibiting its initial *inertia*, will continue to move, seemingly possessing momentum. The analogous relationship between inertia and gravity is shown in Fig. 3.5 (d).

This visual analogy is limited and would only be improved at the risk of further complication, which is unnecessary, for we are now in a position to interpret inertia in what may well be its *true* relationship.

Fig. 3.5 (e) shows a body in free space, the radial lines indicating the creative rays in one plane. It follows that any reluctance the body exhibits to movement is not caused by the body, nor by the rays in themselves, rather the phenomenon is common to both; comparable with the water jet and the plate. As we controlled the source of the jet in order to obtain continual movement of the plate, so by modulation of the creative rays—by an applied force—we bring about an unbalanced condition resulting in continual movement of the body.

We know that for a given mass and a given applied force over a given time, a body will acquire a certain velocity. The theory suggests this velocity is simply the result of the degree of the unbalanced condition, and that inertia is simply the 'resistance' set up by the rays to being modulated. The greater the applied 'force,' the stronger the modulation, and therefore the greater the 'resistance' set up. Furthermore, it will be apparent that only those rays opposite to the applied force are modulated, therefore the body takes on a movement in the direction of that applied force, and what is equally important is the fact that the body will continue to move along the modulated rays (a straight line to our reckoning) according to Newton's First Law, as in (f). The same reasoning can be applied to all aspects of classical mechanics. Even the phenomenon of so-called centrifugal force is now made clear.

Referring to Fig. 3.5 (g), W is a body which has previously received motion by the 'gravitational' or unbalancing effect of X. But it will be observed there will be a disturbance and therefore a corresponding unbalancing with respect to Y. The result of an intermodulation of this kind has the effect identical with the unbalancing effect brought about by X and Y, i.e., a 'new set' of rays are modulated in the direction indicated by Z.

The body now has a new *set* of modulated rays to move along and it will try to do this, but due to increased unbalancing effect of Y, again there is intermodulation.

The process is now continued until either W spirals into Y or, as in the case of a satellite, where the forces—or modulations—are balanced, it will continue to orbit indefinitely.

The whole point is, that because the body is forced continuously to modulate a new set of rays every degree of arc, there is shown a reluctance which we call centrifugal force—or rather the reaction to centripetal force.

Of great importance to the theory is the fact that it is not difficult to modulate the 'etheric' rays. We do so every day of our lives, by moving our bodies or any inanimate object. Avenel has suggested that we probably do the same thing by pressing the button of an electric torch. It is very simple, and we are not aware of it, and more significantly as we shall see it may be just as simple to move a vehicle from our planet's surface other than by the kinetic method of the rocket. At present, in order to lift such a vehicle, we exert a force in the direction we desire it to move, that is, upwards, and in so doing we modulate the very rays which were keeping it in its original state (gravity). Had we an understanding of the 'C' rays we might be able to modulate them by other means—say electrically—and achieve the same result with far less expenditure of energy and discomfort (again we shall have proof of this later on).

We are now in a position to elaborate a little on Avenel's basic theory of gravitation. As stated, it is possible that in some respects my own interpretation and treatment may differ slightly from his, but the difference will be of little fundamental importance to the general hypothesis.

Although the subject of gravity is of prime importance here, extensive development of the theory is beyond the scope of this book. However we can peripherally correlate the known aspects of gravitation to the general concept in the summary which follows.

Inertia -- The 'resistance' offered by the 'C' rays to being modulated. The greater the mass the greater the number of rays to be modulated, therefore the greater the 'resistance' set up, i.e., the greater the force needed.

Velocity -- Rate at which modulations are transferred by resonance to intersecting rays.

Momentum -- Number of rays modulated in direction of motion times the rate of transference of modulation in intersecting rays.

Acceleration -- Unbalanced modulations by a continued interference or applied force.

Gravity -- In the case of two equal masses V and Y, Fig. 4.5 (a), mutual unbalance by interference resulting in continual increase in velocity (acceleration) which is evenly shared. In the case of two unequal masses W and Y, as in (b), greater number of rays being modulated and greater mutual unbalance by interference, therefore continual increase in velocity which is not evenly shared—due to inertia resistance set up in W. If the mass of W is now increased to X, an even greater number of rays are modulated, therefore the higher will be the degree of unbalance and the greater the acceleration, as in Fig. 4.5 (c). If now the mass of Y is doubled as in Fig. 4.5 (d), it will still move towards X with the same acceleration as it did when it was only half the mass, in accordance with Newton's laws.

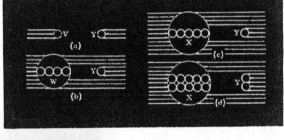

Fig 4.5. Gravitational phenomenon expressed in terms of unity of creation theory.

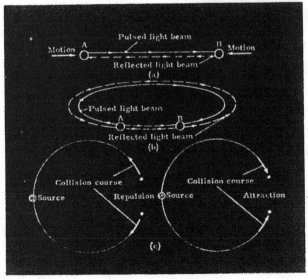

Fig 5.5. Whatever their relative motions may be, two moving bodies are on a collision course when referenced to the source.

Now before we consider some of these facts, one other relevant thought on which to ponder: according to relativistic thinking, if two observers separated by a great distance in a remote space and moving on a collision course, were to send reflected pulses of light at set intervals towards each other, they would be able to measure the diminishing time lag and thereby determine the relative velocity as in Fig. 5.5 (d). Observer A would say the observer B was rushing towards him at velocity V and because of the constancy of the velocity of light, similarly observer B would calculate that A was rushing towards *him* at precisely that speed. Now according to Einstein, Avenel and others, a beam of light would theoretically travel a vast circle, on a cosmological scale, finally bending back on itself. Of course this isn't demonstrably practical, but it serves to illustrate the issue.

Let us imagine that one of the observers, say A, turns 180 degrees with his back towards the approaching observer B, and while in that position sends a stream of timed light pulses directly ahead as before. This time the light will travel a much greater distance in a circle before reflecting back again, but now by calculating the *increasing* time lag, observer A will reckon that he is now *receding* from observer B by precisely the same amount as the previously measured impact speed! In a word, *according to the measuring technique alone,* both bodies are simultaneously approaching and receding from one another; which may seem absurd. [Fig. 5.5 (b)]

Similarly, with attraction and repulsion of two masses, in both these conditions it could be said that all we observe is motion back to the source and that in either instance when viewed from the source, two such repelling or attracting masses are still on collision course with each other, as in (c).

During the elapsing time since the publication of Avenel's Theory of Unity, I have many times found it necessary to emphasise that no matter what point in the matrix may randomly be chosen, common observation forcefully illustrates it represents *a radiant sub-source* in its own right. Also at this juncture, past notification prompts me to underline the point that insofar as the source is both radiant and centripetal, it is reciprocal and therefore self-sustaining.

The Unification of the Big Bang and Steady State Theories?

As is now generally known, today most astronomers accept that the Universe is expanding. This was originally established in 1914 by the American astronomer V. M. Slipher who discovered that spectroscopic analysis revealed a red shift in the coloured photographs of the furthest groups of galaxies. Slipher achieved this with aid of the Doppler effect (discovered in 1842 by J. Doppler, the well-known mathematician and physicist) in which there is an apparent change of frequency and wavelength due to the relative motion of the source of a wave and the observer.

This phenomenon occurs not only with light waves but sound in air as well. In fact, most people are familiar with this latter example through observing a speeding car or a train passing swiftly through a railway station with its whistle blowing. Thus, an approaching source results in an increase of frequency and a shortening of wavelength, and a receding source has the opposite effect. Put more graphically, it is a matter of waves being crowded together on the one hand, and effectively stretched apart on the other. In the case of light we might have a resulting shift towards the red or blue ends of the spectrum. In passing it should be noticed that a very powerful gravitational field can also produce a red shift.

During the 1920s and 1930s, the famous astronomer E. Hubble measured both the distance and the red shift of many galaxies and established what is now known as Hubble's Law; which in effect says that the further the galaxies are from us the faster they are receding. It was also shown that this phenomenon wasn't only relative to the Earth; such recessions might be recorded from other positions in the Universe. In fact, this effect can be readily observed by partially inflating a rubber balloon on which has been painted many bright spots. When the balloon is further inflated it is seen that no matter what particular spot might be randomly chosen nearly all the other spots seem to be departing from *it*. Due to this phenomenon, astrophysicists reasoned that the entire universe must have had its beginnings in some vast cataclysmic explosion which has been estimated as having occurred some 15 billion years ago. Thus the so-called 'Big Bang' cosmological theory was born. It is easy to see that this quickly gained favour due to its biblical interpretation of the times, and remains so to this day.

In 1948, three world-famous cosmologists—Thomas Gold, Herman Bondi, and Sir Fred Hoyle—developed the Steady State theory, which proposes that atoms are continually being created from which all matter, planets, stars and galaxies are formed.

They suggested that coincident with this formation, older stars and galaxies vanish, leaving a constant balance of the material in the physical universe. It is pertinent to note that although both these concepts embrace the Red Shift phenomenon as fact, they leave us somewhat empty-handed as to the eventual fate of these outer galaxies. However, a certain implied paradox manifests. For as it is known that some of the outermost galaxies are receding at nearly half the speed of light, the question arises: Are we to suppose that this is the limit? If not—some lay readers couldn't be blamed for arguing—bearing in mind that relativistic law holds that mass tends towards infinity at the velocity of light, this requires some clarification.

Although this isn't a book specifically about cosmology, clearly its main topic has an inherent close identity with it. Suffice perhaps to add here that some readers will undoubtedly be tempted to make their own deductions on the compatibility of 'dark matter' and 'cosmic strings,' etc., with the general theory.

The Globular Cosmic Matrix

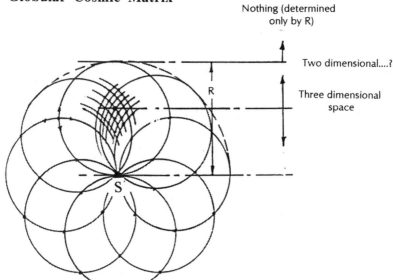

Fig 6.5. The natural gradation from the three dimensional cosmos thro the flat two dimensional hypothetical state, to... 'nothing' according to the theory of unity. Philosophically the source could be identified as the Godhead or more colloquially 'The eye of God'. Note also the close identity with the exponential rate of development as typified in Chapter 2.

I welcome the opportunity to offer a proposition which appears to be more complete than either of the existing theories, which should give it credibility.

On becoming acquainted with the Unity of Creation, there were three more obvious features which became apparent to me; first, the fact that such a system would generate a space matrix, and second, that that would inevitably be of a globular form. But the third most important feature is apt to be overlooked at first until you think about it. That is, the principle elegantly covers an otherwise difficult to imagine change, from a 'state' of 'something' to 'nothing' by creating a three-dimensional grid which changes to two dimensions, or in other words 'nothing,' as amplified in Fig. 6.5.

There can be little doubt that a prevalent feature in the cosmos—as we know it—is that of rotary motion predominantly exhibited both at the microcosmic and the macrocosmic scale; atoms, planets, solar systems and galaxies, most all of these are observed to rotate. I will develop my conception of the globular cosmic matrix further by suggesting that *it* also rotates about the cosmic source, as in Fig. 7.5 (a).

As the entire space-forming matrix is rotating, it follows that all material substances—including so-called dark matter—are also rotating, including atoms, planets, stars, solar systems and, not least, *galaxies*. One of my specialities in the aerospace establishment had been aerodynamic fans, particularly centrifugal fans. So I have the advantage to know that in such a rotating system, particles which are radially flung out at the centre meet ever-increasing rotational speeds the further out they become, as in (b). The further out the particles reach, the greater is the localised *dispersion between them*: similar to the previously described rubber balloon example, and galaxies!

No doubt some readers will anticipate me, but for the record I now suggest that exactly in keeping with the steady state theory, particles are formed by the transformation of pure energy from the cosmic source. Due to the natural electro/gravitational clustering, masses become more dense and due to the resulting ever-increasing centrifugal force, tend to migrate toward the outer regions we call space. The more dense the conglomerants become, the further out they spread in the form of planets, stars and galactic clusters. In keeping with the rotating centrifugal fan phenomenon, the outer galaxies will reach the highest speeds and thereby exhibit the red shift to watching astronomers as in Fig. 7.5 (c) and (d).

So far, all this is in agreement with both the steady state and the big bang theory, but what of the relativistic velocity of light barrier previously mentioned? I suggest that by the time the outer galaxies enter these distances their attendant increase in mass is proportionally compensated by the ever-shrinking

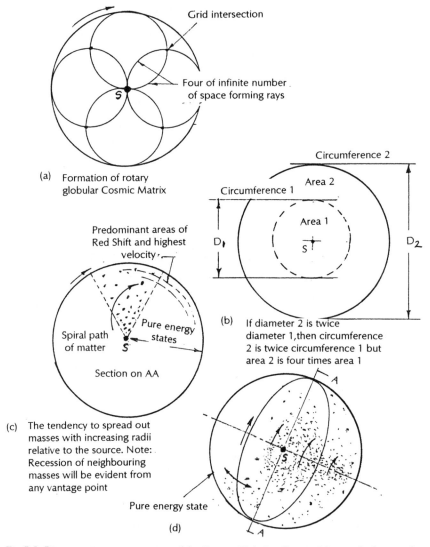

(a) Formation of rotary globular Cosmic Matrix

Grid intersection

Four of infinite number of space forming rays

Circumference 2

Circumference 1

Area 2

Area 1

D_1

D_2

(b) If diameter 2 is twice diameter 1, then circumference 2 is twice circumference 1 but area 2 is four times area 1

Predominant areas of Red Shift and highest velocity

Spiral path of matter

Pure energy states

Section on AA

(c) The tendency to spread out masses with increasing radii relative to the source. Note: Recession of neighbouring masses will be evident from any vantage point

Pure energy state

(d)

Fig 7.5. Diagramatic representation of the Rotary Globular Cosmic Matrix which according to present reckoning might have a diameter of over 16 Billion Light Years. The author's hypothesis would appear to reconcile both the Red Shift and the Steady State Theories. Note: For clarity a fixed plane of Globular Rotation has been shown but this may also be rotating.

third dimension until they reach the outer limits of two dimensions and extinction back into the pure energy state from which they were originally formed. I further suggest that this part of the process is continuous—exactly as predicted by Sir Fred Hoyle, Einstein and others. Surely such close overall agreement lends substantial support to the Theory of Unity.

The Formation and Interaction of Modulated Wave/Particles

During the past four decades or so there have been few major developments in astrophysics which have persuaded me that neither the Theory of Unity nor my contributions to it are significantly at fault. Therefore not only may it be interesting to the reader, but necessary, to include my recent additions to it here.

It will be seen that further examination of the broad concept would suggest that with suitable amplification, we can go further towards unifying the general theory by including the formation of the individual subatomic particles,

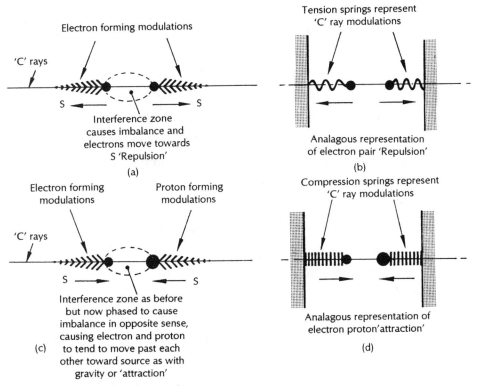

Fig 8.5. Unbalance or reduced symmetry produces electrostatic attraction or repulsion effects mechanically represented.

beginning with the electron.

Fig. 8.5 (a) represents one of many 'C' rays in one plane which have been locally modulated at a certain frequency and coincident point to form an electron. If another adjacent electron is formed whose extension modulations coincide with those of the first electron, we can imagine an interference occurring which, akin to common wave principles, can either be constructive or destructive, depending on the phasing. In the case of electrons we might justifiably assume that an imbalance is created, as in the case of gravity, only in the opposite sense, as the electrons—now behaving as rays—move away from one another in their travel back to source. Fig. 8.5 (b) is an analogous representation of this in which the modulations are simulated by tension springs.

Fig. 8.5 (c) represents an electron-proton pair. In this, the proton-forming modulations are phase-shifted by 180 degrees, which causes an interference between the proton- and electron-forming modulations, in which case the protons and electrons move *toward* each other on their travel back to source, as in the case of 'attraction' or gravity. Fig. 8.5 (d) is an analogous representation of this in which the modulations are now simulated by compression springs. Note, this last analogy is penalised somewhat by the fact that the inverse square law effect is reversed, however the effective similarity is sound enough.

For the sake of clarity, there is one important factor which has been neglected from the foregoing generalised portrayal, that of gravity, for over and above the electrical strong forces, the weak force of gravity still resides and this must now be included.

Fig. 9.5 (a) represents two free electrons in close proximity indicating the interference zone between them. For convenience it is assumed the waves in this zone are in such a phase which produces modulation imbalance to render electrons seemingly repellent to one another. We can now imagine these free electrons are but two of billions in two comparatively massive bodies which tend to be moving away from one another in their travel back to source. However the gravity-forming modulations—having different wave characteristics to those of the individual particle waves—are not one bit nullified and therefore still exert an apparent 'attraction' force in opposition to the electrons' repulsion.

Fig. 9.5 (b), (c), and (d) are self-explanatory arrangements which portray the interplay between the so-called strong electrical and the weak gravitational forces as they would be in contention in some normal situations.

Later on it will be shown that the frequency of the particle-forming modulations are probably not of the same order, and it should be borne in mind that Avenel suggested the basic 'C' rays might be alternating. Further, the particle-forming modulations take on some of the carrier wave characteristics. This, together with the complex omnidirectional nature of the C rays, renders it difficult (to say the least) to meaningfully portray the concept more graphically.

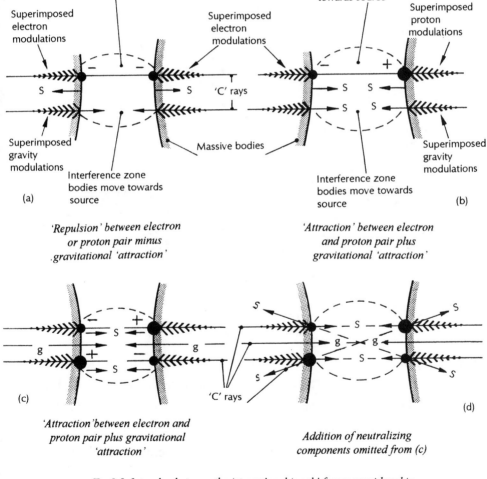

Fig 9.5. Interplay between the 'strong' and 'weak' forces considered in terms of the theory of unity.

But it is interesting to speculate that the foregoing combined effects may throw some additional light on the puzzling 'Waveicle' dual nature of apparently orbiting electrons!

Magnetism and Induced Currents

As stated earlier, when atom-forming modulations are displaced, the modulations at right angles to this motion are transferred by resonance to other 'C' rays in all planes around them, much the same as radiating expanding ripples on water. The same may be said of the electron.

Although full development of the principle is beyond the scope of this book, a cursory consideration of what has been discussed so far will reveal a

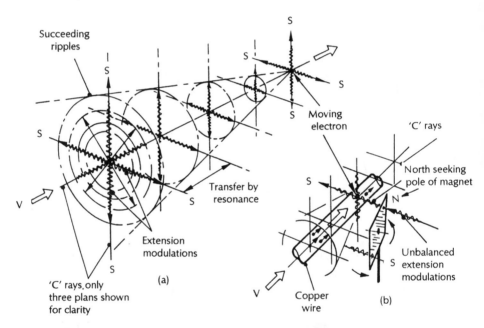

Fig 10.5. Atoms in a magnetized body unbalanced due to a moving electron.

more complete structural assessment which includes magnetic phenomena, that is, as I see it in terms of the Theory of Unity. For instance, as shown in Fig. 10.5 (a) the progressive transference of the extension modulations from one 'C' ray to another in association with the corresponding displacement V, produces a conical-shaped 'wake,' much the same as does an aircraft through the air or a boat in water.

Developing this analogy further suggests that just as the aircraft generates *an advancing* shock wave in air and the boat an advancing bow wave in water, then in terms of the Theory of Unity we shouldn't be surprised to learn that a moving electron does exactly the same thing in a structured 'ether.' (B) represents one of many electrons together with the associated extension modulations moving along a conducting wire. In this instance, the conically expanding modulations cut across a nearby magnetised body (say in the form of an ordinary compass needle) in which some of the atoms are magnetically plane polarised to a degree. Depending on the relative wave phasing, some of the extension modulations are unbalanced even more and they tend to move towards S; in this instance *towards* the travelling electrons exactly in accordance with well-known, but not, I venture to add, fully understood electromagnetic laws. For those who may find it useful, the following is another of my hydraulic analogies which in the past had proved so useful to me.

In Fig. 11.5 (a) we adopt the above boat analogy showing a vessel producing waves in the water as it proceeds along a canal. At the point X on the canal bank stands an observer watching the approaching waves dissipating their energy on the bank before him. If we imagine that the observer was holding a transparent pipe dipped into the water so that he could see the water level, he would of course notice that as each wave broke onto the pipe some of the energy would cause an increase in the water level. If this energy were sufficiently energetic, some water would shoot out of the top end and he would get very wet! However suppose he now lengthened the pipe and curved it over to the opposite bank, then this discharge could be deposited there. If the observer wished it, he could make this discharge do work and, say, turn a turbine.

We come to the limit of this part of the analogy by proposing that the observer causes the curved pipe to descend into the water at the opposite bank, in which case clearly there would be a reciprocal action which would cancel out the effect. In (b) the analogy is extended a little further by imagining the curved pipe to be completed into a circle so that the lower half is submerged. Next, the interior of the pipe is fitted with a 'winding staircase' type spiral. At points A and B (level with the water) holes are formed in the pipe facing the oncoming waves from the boat, designated V_1 and V_2. The water V_1 entering the orifice at A will begin to rotate up or down the spiral depending on the helical angle. In

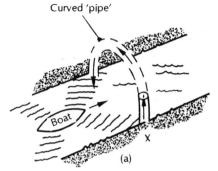

Curved 'pipe'

(a)

First stage of an induced current analogy.

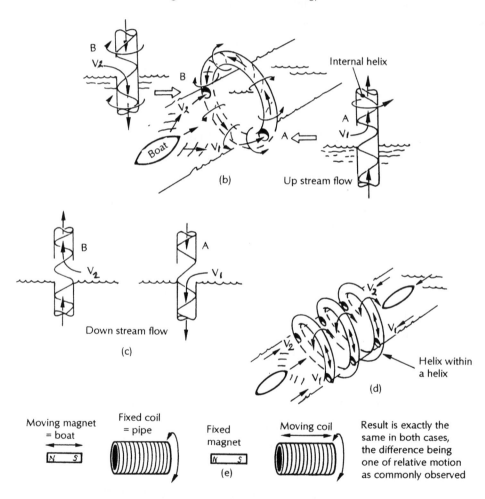

(b)

Up stream flow

Internal helix

Down stream flow

(c)

Helix within
a helix

(d)

Moving magnet
= boat

Fixed coil
= pipe

Fixed
magnet

Moving coil

Result is exactly the
same in both cases,
the difference being
one of relative motion
as commonly observed

(e)

Fig 11.5. This analogy of induced EM currents is more supportive of an energetic expanding wave phenomenon rather than the presently held static field concept. It is emphasized that the water in the above analogy should be interpreted as a thin slice of an otherwise cylindrical shaped field.

the diagram it is assumed the rotation to be anticlockwise and the water motion to be rising as indicated. It will be noticed that due to the curvature of the pipe, the water on the far side will now be seen to be rotating *clockwise* and *descending* toward orifice B where it will be joined by water V_2. Thus the action is no longer cancelled out as in (a) but amplified to cause a complete and continuous cycle, so long as there are oncoming waves from the boat.

Now this doesn't necessarily imply that there has to be a one-way traffic for the action to take place, as a study of (c) will reveal. This shows that by the simple expedient of placing adjacent and opposite holes in the pipe a returning vessel would produce exactly the same action *but reversed*. For on entering the orifice at A the water would now be turning clockwise and *descending*, while at B it would be turning anticlockwise and *ascending*. In other words the motion is reciprocal depending entirely on which way the boat—or waves—were travelling, up stream or down.

Fig. 11.5 (d) takes the analogy to its conclusion by imagining the single loop to be extended into a spiral, into which the little boat is caused to move in and out and we have an exact duplication of the common magnet and coil-induced current process as indicated in (e).

It is important to notice that the authenticity or otherwise of the foregoing analogy depends entirely on the fact that the inverse nature of the principle involved is brought about by nothing but a *change in angular orientation* of the components, in exactly the same manner as the previous rotating cylinders analogy of magnetic poles shown in Chapter 3. These, and other analogies in this book, are but a few; there are many more which all help to reinforce the notion of expanding waves rather than the currently held static field inclination. They represent the helping props I dreamed up as a pioneering novice, and for some, I believe they may well prove to be useful today.

A More Convenient Approach to a Complex Subject

During the past four decades or so I have become acutely aware of the difficulty most people have in visualising something so abstract as the interplay of an intricate cosmic ray system. In order to accommodate that I have perforce, from time to time, varied the descriptive analogous approach according to the prevailing circumstances. For instance, while I have tried to show that it is true the foregoing interference set up between atom-forming waves could in some ways be compared to a ray of light passing through a transparent substance—thereby being 'absorbed' and transferring some of its latent energy into heat—the idea of a ray passing through a *massive* body can be extremely difficult for some to visualise, even though on a day-to-day basis we accept the fact that radio waves pass through the walls of our homes and many other physical objects before reaching our radio sets. Having established some of the arguments for the basic theory, I have usually found the middle course analogous approach to be a satisfactory compromise. Although the indications are representative rather than definitive, I propose it will prove to be a less painful and generally more expedient procedure for us to adopt from now on.

Crookes' Radiometer as a Step Analogy to Gravity

On many occasions I have found Crookes' Radiometer to be a time-honoured visual aid, which will help to convey the general idea here. It was invented by Sir William Crookes in the last century, who interestingly enough believed the device bore evidence of direct mechanical effect of radiation. It consists of a glass globe from which the air has been partially exhausted and a small centrally placed turbine-type wheel with four small paddle-like blades which spins on a low friction needle point bearing. Each blade is finished silver on one side and black on the other and so arranged on the support arms that the rotor will turn when sunlight is beamed towards it [Fig. 12.5].

Originally, soot was used to blacken one side of the blades and Einstein

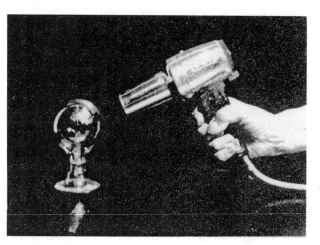

Fig 12.5. Crookes Radiometer as a step analogy of 'gravity as a deficit' Note: Two opposing waves do not cancel each other out as would two opposing jets of air or water. In the above a hot air gun is used to supply the required heat.

suggested that the heat contained in the light caused tiny particles of it to be ejected, providing a rocket-type reaction on one side of the rotor, causing it to revolve. However this became questionable when the blades were painted black with other less easily removed substances.

The phenomenon was also attributed to the light being reflected off the silver side and absorbed by the dark side, but this would require the rotor to spin in the opposite direction than it does. This can easily be established by alternately blanking off first one half of the glass bulb, then the other, while light—or heat—is directed towards it. In fact, while doing this it will be found that the rotor will turn even faster due to the fact that the counter-rotative effect produced on the silver side has been negated.

That there *is* a degree of corpuscular reaction due to the reflection of light has long been known and was attested in the 1960s when the huge inflated sphere of aluminium foil carried by the Echo 1 satellite caused a severe orbital deviation round the Earth due to the 'pressure' of sunlight.

Today, it is widely believed that the Crookes Radiometer effect is caused by heat radiation which excites some of the remaining air molecules in the glass globe, which therefore bounce off the rotor blades—more on one side than the other—thereby producing the rotation. There are some obvious objections to this, but we needn't deal further with them here; suffice to say that this little device does help to graphically illustrate that the turning effort on the rotor is produced by nothing more than heat radiation—or waves in the electromagnetic spectrum—*passing through* the glass molecules of the globe. The effect is even more marked if the radiant heat is beamed at the rotor from all sides, i.e., through 360 degrees.

We can now take our analogy a stage further by imagining the experiment to be repeated far out in space and the rotor blades to be removed from the rotor arms and left weightless inside the globe. Furthermore, if they could be oriented to present all the silver sides towards the centre of the globe and then subjected to an omnidirectional light or heat source, they would conceivably all tend to cluster together in the centre.

The next step in this most important analogy is to imagine that the foregoing little turbine blades, or wings, could be further modified by piercing

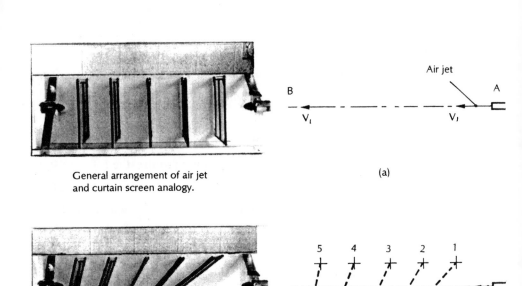

General arrangement of air jet
and curtain screen analogy.

(a)

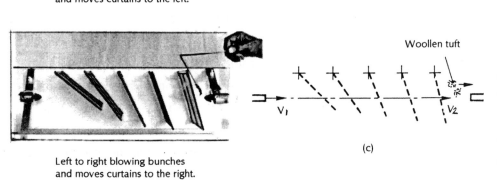

Right to left blowing bunches
and moves curtains to the left.

(b)

Weighted net curtains

Left to right blowing bunches
and moves curtains to the right.

Woollen tuft

(c)

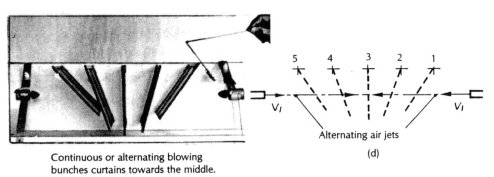

Continuous or alternating blowing
bunches curtains towards the middle.

Alternating air jets

(d)

Fig 13.5. Air Jet and Curtain Screen Analogy of gravity as a deficit.

them with many holes through which some of the radiant energy passes ineffectively. Naturally, the degree of movement would be lessened but the clustering tendency would still be there. This fact can be illustrated suitably by the next step analogy and with it my assertion that gravity may be conveniently likened to *a kind of 'deficit'* rather than *a 'pull.'*

Gravity as a Deficit

Fig. 13.5 (a) depicts a jet of air issuing from a nozzle at point A moving towards B, at which position its velocity V_1—or energy—is measurably much the same as it is at A.

Fig. 13.5 (b) shows a series of four weighted open-weave curtains—or screens—hanging at intervals in the path of the air jet. When the jet commences it first passes through screen 1, giving up some of its energy in so doing and displaces the screen considerably. Next it passes on to screen 2, but now because it is moving slower, displaces this screen less than screen 1, and so on through screens 3, 4 and 5. After exiting at screen 5 the jet velocity is considerably reduced to V_2, but notice that the screens have been displaced by correspondingly lesser amounts, thereby catching up or collecting together (as with gravity). In this analogy the jet represents 'rays' in one direction and plane only, while the screens represent particles of matter. Of course, two opposing jets have a more representative effect, blowing the screens to the centre as shown in (c) and (d). The limitation of this analogy is that it is impossible to accurately represent the case truly, since the two jets (rays) should be diametrically opposite. I found this can be accommodated by making the jets alternate one after the other and the result would be exactly the same.

Fig. 14.5 (a) takes this principle a stage further showing only one space 'ray'—representing trillions—approaching a massive body such as the earth, 'passing through' it, thus tending to squeeze all the atoms together as in the above analogy—giving up some of its energy in the process—before leaving the earth at the other side 'beneath,' say, an adult male situated there. Due to its relatively gargantuan mass the earth wouldn't move perceptibly (as we saw in Fig. 4.5 (c), above) but the 'diluted rays' still have colossal energy to move the individual away—or up—from the surface, but for the fact that the opposing *'undiluted rays'* are pressing his body down more powerfully by an amount—in this

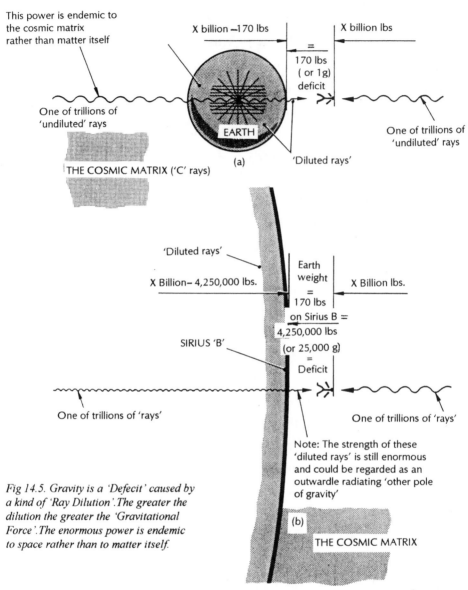

This power is endemic to
the cosmic matrix
rather than matter itself

X billion –170 lbs

X billion lbs

=
170 lbs
(or 1g)
deficit

One of trillions of
'undiluted' rays

EARTH

One of trillions of
'undiluted' rays

THE COSMIC MATRIX ('C' rays)

(a)

'Diluted rays'

'Diluted rays'

X Billion– 4,250,000 lbs.

Earth
weight
=
170 lbs

X Billion lbs.

on Sirius B =
4,250,000 lbs
(or 25,000 g)
=
Deficit

SIRIUS 'B'

One of trillions of 'rays'

One of trillions of 'rays'

Note: The strength of these
'diluted rays' is still enormous
and could be regarded as an
outwardle radiating 'other pole
of gravity'

*Fig 14.5. Gravity is a 'Defecit' caused by
a kind of 'Ray Dilution'. The greater the
dilution the greater the 'Gravitational
Force'. The enormous power is endemic
to space rather than to matter itself.*

(b)

THE COSMIC MATRIX

case—of approximately 170 lbs. I believe it is this 'deficit' between the opposing 'rays' that we call gravity. In a word, there is more of it 'pushing' us down than there is 'pushing' us up. The reader will see the significance of this is awe-inspiring indeed, as follows.

We have seen that due to the mass of the earth, this deficit (or 1g) amounts to approximately 170 lbs. in earth units, therefore it can easily be calculated what the force on the individual would be if the earth's density was that of the dwarf star Sirius B at no less than 25,000 gs. As a certainty, he would be very flat! [Fig. 14.5 (b) and Fig. 15.5]

Again this would be due to the gravitational 'deficit' which may help to convey the fact that if the energy of the rays we cannot see nor feel could be assessed in terms of pressure, it would be enormous! Thus, far from being empty, space is a vast reservoir of unimaginable power and what we call gravity does not exist in the sense of a 'pull,' neither does it originate from *within* matter (as is presently held at even advanced levels in some areas of contemporary physics), matter is entirely dependent upon and subservient to *it*.

Stellar and Planetary Heating

Although cosmology is not a prerequisite of this book, the developed theory of vehicular propulsion and energy is so inextricably interwoven with it, that it is important not to totally omit supportive arguments. An obvious example being: If we accept that energy is 'absorbed' from such rays on interaction with a massive body, then such absorbed energy must be transferred into another form. Without too much stretch of the imagination, we might suspect that such a transference would logically be expressed as heat. Moreover, we would suspect that such heating would tend to be concentrated at the centre of large globular bodies, increasing in intensity with increase in size and mass, which of course is true of all celestial bodies from natural satellites to planets and stars. Granted, the currently accepted temperature rise with increase in pressure action accounts for most of the heat in the formation of planets, but the foregoing conclusions are not at variance with this. Additionally, the author's gravity as a deficit hypothesis would appear to answer the oft-repeated question, "Why is gravity monopolar?" The answer being, "It all depends on how you look at it." For if we could subdue or neutralise the 'downward' pressure on our heads the 'diluted' radiant energy being emitted 'upwards' beneath our feet—in that sense—can become a seemingly gravitational 'repulsive' force (or 'antigravity') as you wish. In the following chapters some directives are offered as to how this might be achieved.

Attraction and Repulsion Unified?

So gravity as a kind of deficit proposition inherently reconciles the apparent bipolar anomaly in nature which according to Occam's Razor's ruling

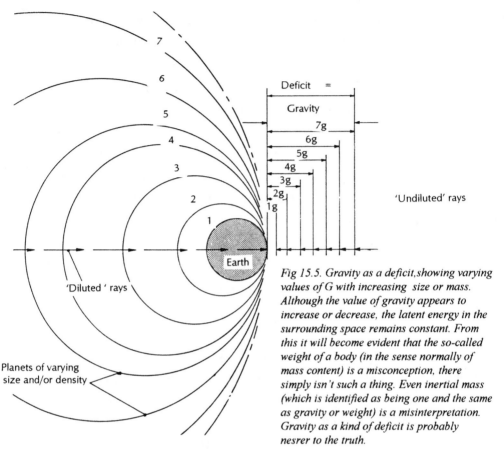

Deficit = Gravity

'Undiluted' rays

'Diluted' rays

Earth

Planets of varying size and/or density

Fig 15.5. Gravity as a deficit, showing varying values of G with increasing size or mass. Although the value of gravity appears to increase or decrease, the latent energy in the surrounding space remains constant. From this it will become evident that the so-called weight of a body (in the sense normally of mass content) is a misconception, there simply isn't such a thing. Even inertial mass (which is identified as being one and the same as gravity or weight) is a misinterpretation. Gravity as a kind of deficit is probably nesrer to the truth.

lends further credence to the general theory. This can be further illustrated by combining the Crookes Radiometer experiment, the air jet and screens analogy and that shown in Fig. 16.5. Thus by virtue of the fact that rays of light are involved instead of air jets, we are taking a step closer to understanding the basic principle involved.

This almost self-explanatory analogy portrays three sets of screens. The uppermost set (A) are clear glass, the middle set (B) opaque glass, and the lower set (C) are plane-polarised glass. For convenience these are shown larger but of course they could easily be the size of the small vanes in the Crookes Radiometer, in which they would have to be independently supported. The experiment is as follows.

(1) Two opposing beams of light—instead of the previous air jets—are projected onto the transparent screens (A) which, according to the foregoing

drag deficit, would theoretically move towards one another; this represents gravity in the analogy.

(2) The second set of screens (B) being opaque, when the two beams of light are projected onto them as before, due to the increased drag will tend to move together as in (1) but even more so. This represents magnetic and

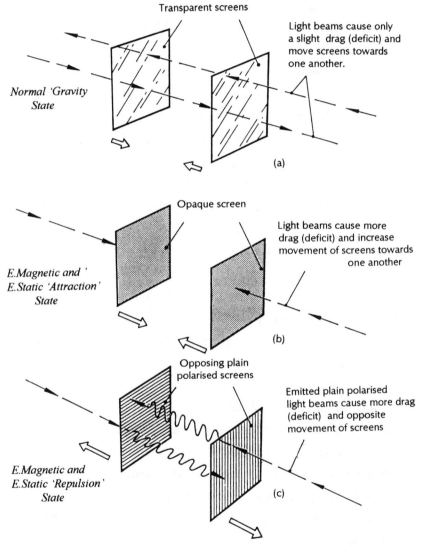

Transparent screens

Light beams cause only a slight drag (deficit) and move screens towards one another.

Normal 'Gravity State

(a)

Opaque screen

Light beams cause more drag (deficit) and increase movement of screens towards one another

E.Magnetic and ' E.Static 'Attraction' State

(b)

Opposing plain polarised screens

Emitted plain polarised light beams cause more drag (deficit) and opposite movement of screens

E.Magnetic and E.Static 'Repulsion' State

(c)

Fig 16.5. Step analogy of the main natural effects in which directly opposing beams of light represent the Cosmic Rays and the screens represent two isolated bodies in space. Note: The author intends this as purely hypothetical as the pressure of light at this scale would be negligible. Also the reader will realise that such reasoning places us at the very threshold of more meaningful and searching experimentation.

electrostatic 'attraction.'

(3) The opaque screens are now replaced by the set of polarised ones (C) and the experiment repeated. However in this case, due to the relatively free transmission of the light beams through the first screen and its total absorption by the second screen, more drag is experienced by it and vice versa, thus the screens would move apart. This represents the so-called other pole in magnetic and electrostatic repulsion. When considering this analogy it will be profitable to bear in mind that until fused into glass, its basic constituents (sand metallic oxides, etc.) *are normally opaque to light waves.* [It cannot be overstated that when the foregoing observations are correlated with the Theory of Unity we begin to recognise the emergence of support more in favour of Hoyle, Bondi and Gold's steady state cosmological theory, rather than that of the Big Bang.]

Although this part of the hypothesis has been simplified and can only be more accurately discussed by complication, the basic ideas for further speculation are there. If it can be shown there *is* merit in the basic theory, then it will help the reader in following the possibility of employing other means for generating energy and propulsion later on.

Summary

As I shall be closing this work with some—by present-day standards—rather extraordinary ramifications, it will be necessary for the reader to have a certain rapport with the principles involved thus far. The following summary may help to consolidate this.

1.   The theory holds that 'attraction' and 'repulsion' forces concerning matter, *do not* originate or reside *within* matter.

2.   Rather the phenomenon that causes these effects resides in what we call 'space' and that 'attraction' and 'repulsion' are merely the effects of this energy being unbalanced or changed on interacting with 'matter.'

3.   This explains why the degree of attraction becomes greater as the bulk and density of matter increase, i.e., the more massive it is the greater the 'dilution' or 'absorption' of the rays.

4.   Accepting (2) that the 'power' of so-called space itself is responsible for gravitational, magnetic and electrostatic field phenomena, we can get an idea of the colossal energy in space by noting that a 'weight' of, say, 200 pounds on the

earth would become about 5 million pounds on the dwarf star Sirius B—while on a neutron star this meagre weight would be increased to a thousand billion times that. By this reckoning we may interpret the colossal power of space *to be infinite!*

[As stated in the beginning of this chapter, in 1954 when Avenel first published his "Unity of Creation Theory" he gave a tentative estimate of the wavelength of the creative rays as 10^{-13}cm. More recently Dr. John A. Wheeler of Princeton has calculated the wavelength of the 'ether' as 10^{-32}cm. This is roughly sixty octaves above the limit of the known electromagnetic spectrum, which is based on the diameter of an electron (approximately 5×10^{-13}cm). Wheeler calculated an incredibly high energy density, from 10^{94} to around 10^{127} watt-seconds per cubic centimetre, which in more graphical terms means that if the potential energy of one cubic inch of Avenel's space were converted to electricity it could power an average city. Not only has the 'ether' wavelength been estimated, but it has been given a tentative density approximating to 10^{-20} that of water.]

When considering all that has been said so far, it should be borne in mind that, of the observable forces residing in nature, gravity appears to be the most inferior, just how much so was indicated in the last chapter. However it must also be remembered that within the context of this theory, it is *ray absorption* or imbalance that produces the foregoing 'deficit' which we presently call 'gravitic,' 'electrostatic' and 'magnetic' forces.

In future years we may find it normal—and the author feels certainly more correct—to refer to the term 'gravity deficit' rather than 'pull of gravity' as is commonly used in textbooks and—not least—conventionally by all leading cosmologists of the day.

The Cosmic Matrix (Comat)

Although for convenience I have used the descriptive term Cosmic Matrix as in Fig. 14.5 and elsewhere, perhaps the acronym 'Comat' will suffice from now on. In considering the propositions set out in this chapter we are in good company, for Sir Isaac Newton is quoted as having said:

> For it is well known that bodies act upon one another by the attractions of gravity, magnetism and electricity; and these instances

show the tenour and course of Nature and make it not improbable that there may be more attractive powers than these. For nature is very consonant and comfortable to herself.

Naturally it is anticipated that many physicists will be unable to agree with the foregoing hypothesis, in which case the author hastens to point out, within the present framework, overall agreement at this time is unnecessary, particularly if we are mindful of the fact that *based solely* on the original early supposition of 'magnetic rays' or 'lines of force,' science has been usefully exploiting electromagnetism for many years now. What I am proposing here is that there are sound reasons to suggest that we extend the range of that basic thinking.

A Broader Unified Field Theory?

By now it will be apparent that the Theory of Unity not only embraces the so-called weak force of gravity and the strong force of magnetism, but logically it also includes the strong electrostatic force and any apparent difference between such phenomena is only a function of wavelength and/or frequency. In Chapter 7, I offer an explanation for the phenomena of attraction and repulsion as being one identical action which is true or false depending only on the authenticity of the basic theory.

It is important for the reader to understand that any realistic appraisal of a cosmological theory should inherently include as many of the natural phenomena as possible; including not only the known and well understood, but also the little known and less understood. This is a stand I take throughout this work and it is necessary that from time to time we must remind ourselves of it. In the following section I have attempted to do just that by extending the range of this report to include both normal phenomena and some aspects of the currently so-called paranormal. All of which, I submit, should be taken into consideration when evaluating the final conclusions, particularly in terms of the Theory of Unity. For some, the inclusion of some of the material may appear to be somewhat incongruous, but I must emphasise, a true evaluation of the remainder of this work is dependent on it, for just as chemists study new fuels for rocket motors, we shall, after all, be examining some of the 'constituents' of the kind of 'fuel' which will assuredly send our future spaceships to the stars and back.

Footnote

At the time of going to press, there have been spurious statements in the news media concerning the work of cosmologists who are apparently searching for evidence of so-called 'Dark Matter' which they suspect "holds the Universe together"! Currently some of this work is being conducted in Europe's deepest mine at Balby Potash Mine, Cleveland, Yorkshire, UK.

1 See Appendix 2.

PART THREE

INCLUSIVE ATTRIBUTES
OF THE COSMIC MATRIX

6

PARANORMAL ATTRIBUTES
Of the 'Spooky' Kind

A Cosmic Cake

Over the years student physicists the world over have become acquainted with the attempts by Einstein and his successors to find one unifying field theory; one basic idea which will successfully correlate both the so-called weak and strong forces in nature. Respectfully, I suggest that this will never be achieved until additional observable field phenomena are included. For physicists to continue to do otherwise is surely tantamount to Granny preparing a Christmas cake with only one or two ingredients! Likewise it seems to me that most extravaganza concerning subatomic excursions into theoretical attempts to control gravity by 'plasma jets,' etc., is comparable to trying to design and build steam engines without having the slightest notion as to the constituent basis of water; but statements by world-leading scientists of the calibre of Dr. Leon Lederman (quoted in Chapter 1), i.e., "We are drowning in theoretical possibilities not based on a single salutary fact," should be supportive enough!

In this and the following chapter we shall be merely scratching the surface of what after all could be one of the most important, yet disregarded ingredients of them all. It is accepted that some of the information will provoke scepticism among some inexperienced readers, so before we proceed further it may be as well to include a word or two about the self-confessed sceptic.

The 'Authority' of the Sceptics

No matter how journalistically proficient some of the sceptics might be concerning paranormal phenomena, the truth is that they should be dismissed as of little consequence, for without exception they are totally inexperienced as far as the subject is concerned; the very fact that they *are* sceptics automatically qualifies this judgment, for if they had *really* enquired into these matters they certainly wouldn't be sceptics! It's rather like an untrained person pronouncing on advanced aerodynamics, a subject about which they haven't a clue. I would go further by restating to those lay persons—who would otherwise seek opinion on the paranormal from experts in other more conventional disciplines—that such 'authorities' should be casually indulged and not taken seriously.

However, neither is it only establishment diehard sceptics we must endure, for among the predictable increasing numbers of Ufologists, for instance, can be recognised those few—who no doubt due to mental exhaustion by their sadly inadequate efforts to explain technologically engendered side effects they designate as high 'strangeness'—try to persuade the rest of us that *all* the UFO phenomena together with the paranormal kind (including poltergeists, etc.) are induced by the 'collective unconscious' or are electromagnetically induced, sometimes shared, hallucinations! What they seem unable to understand is that with an advanced technology, there inevitably *will be* a 'strangeness factor,' a fact that in no better way was expressed by the late Dr. J. Alan Hynek when he said, "A twenty-first century technology will be indistinguishable from twentieth century magic!"

Obviously, I am not saying that *all* the strangeness effects are induced by a technological process—some of it *is* deliberately induced at the conscious level for other purposes. But I suggest it is a grave mistake to accredit *all* the phenomena to one or the other cause.

One of the most transparent arguments employed by scoffers of the paranormal is the claim that protagonists of the subject rely chiefly on old spiritualist stories of the last century. This is not true; many people—including this author—have experienced spontaneous effects which cannot be explained conventionally and where trickery is quite out of the question. Granted, few would disagree with sceptics when they say, "If there are scientists who are gullible and easily led by tricksters, then they shouldn't be scientists." This convenient objection should by no means be accepted as a general trend. I, for one, have always been cautious in my enquiries to a degree bordering on paranoia, and with the same sense of exactitude one adopts in an engineering workshop. This exactitude enables aeroplanes and rockets to fly, which they certainly wouldn't if one adopted the easy casual attitude painted by the sceptics.

The only true part of their argument lies in the fact that more cases of remarkable paranormal phenomena occurred in the last century than today, and for several obvious reasons when you think about it. In the first place, many people were inclined to be more religious in the days before World War I, while many in the middle class certainly had less to do. So the social climate was ripe for spiritualistic inclination, which was accentuated after the war. Spiritualist churches started to spring up and more people witnessed phenomena demonstrated by mediums. This soon developed into privately run materialisation 'circles' in America and Europe, no doubt some of which were fraudulently based, but a good many were quite genuine.

Between World Wars I and II, with the introduction of radio broadcasting and consequent burgeoning dance and music halls, etc., people had other things to do. Fewer spiritualist circles were formed and consequently the decline of good mediums. Today their numbers are few, but here and there all over the world private circles are still formed. During my life I have known people to do exactly that, more often than not out of curiosity—which I wouldn't say is the best motive—and some have been amazed and quite unprepared for the phenomena that occurred. *They* certainly weren't fooling themselves!

Predictably, some scientists who *have* studied the paranormal naturally find it difficult to reconcile with modern physics, notably Professor John Taylor of Kings College, London[1] (with whom I shared a lecture platform in the 1970s). He and others have measured the brain waves of people in attempts to correlate brain wave patterns with such physical phenomena as levitation, etc., presumably in the hope that the extremely weak electromagnetic fields might prove to be *directly* responsible, despite the fact that in some cases objects as massive and heavy as fully laden wardrobes have been *lifted* and moved *several* feet! Whether or not the experiments are conducted in the hope that such weak fields will lead to *something else* is speculative, but the fact remains that prior consideration of the cosmic matrix theory might be of considerable assistance. In other words, both otherwise well-intentioned physicists *and* some paranormal researchers may have only part of the story. However, it must also be said that today a steadily increasing number of scientists are becoming interested in paranormal phenomena, no doubt in part due to the wider media coverage.

It must be emphasised that we are now going to take a brief look at some seemingly strange facets of nature, which are only strange when they are considered in isolation. But the cosmic matrix manifests in many guises and only by viewing these as a whole will we be able to identify and eventually usefully exploit this vast resource. In view of the somewhat 'otherworldly' aspects of the phenomena, it must be reemphasised this author is emphatically not a self-deluded psychic, rather an unbiased enquirer who has personally had the good fortune to experience both 'normal' and the 'paranormal' sides of life.

Some Aspects of the Primary Energy State

Despite the currently accepted meaning of the term primary energy in modern physics, in order to avoid any confusion, the distinction should be made that I regard the space matrix as truly constituting the *primary energy state* and its physical counterpart (matter) as the *secondary energy state*.

It is important to recognise that it is somewhere between these two levels or states that we find the *known* electromagnetic spectrum; it is also the domain of many unknown dimensions and forces. Indeed, we may find that eventual control of gravity will require recognition of such forces over and above the purely electromotive variety.

Although I have emphasised the existence of far more profusely informative sources for paranormal activity, and in particular levitation, due to the pivotal nature of this one aspect to the rest of the work, I have considered it desirable to be more liberal with some inclusive reports. As other kinds of paranormal phenomena can be explained in terms of the general theory, without straying too far out of context, I feel some of the evidence for this should also be included in this section, in a generally supportive role.

Although the realm of paranormal phenomena isn't central to the main issue here, it is nonetheless vitally important that it should be listed as a natural function within the matrix. To some, its inclusion will no doubt seem strange; that it should be segregated into two broad factors may seem even more so. But the fact of the matter is that although the origins of the phenomena are no doubt common, I have noticed there are two broad bands of a discernible difference among it all.

This first half of the broad spectrum is recognisable by the fact that it seems to be of the mental variety in which the phenomenon depends on a direct implementation of the mind of the participant. There are other examples, but the chief of these can be listed as follows:

1. Precognition (Time travel? Witnessing events before they occur)
2. Telepathy (Mind to mind contact between sender and recipient)
3. Clairaudience (Seemingly audible voice transmissions)
4. Psychometry (Mental discernment of past events by link-up with the auric field which permeates all natural objects)

5. Remote Viewing (Similar to Item 4, but visual information of current affairs are discerned)

6. Apparitions (Visual discernment of people both alive or dead, which is frequently selective among witnesses).

As our space here is limited, not all of these mental phenomena can be treated at length and in the following examples, I have once again drawn on my own experience as being the instances I know best, and which (as a devoted unbiased seeker of natural phenomena—of any kind) I can vouch for.

Precognition —Are Some People Time-Travellers?

At the beginning of Chapter 2 we saw a statistical method by which certain future trends might be reasonably predicted and to the best of our knowledge that is the only instance of *scientific* forecast available. So-called precognition, on the other hand, has been around since the beginning of recorded history, and it is interesting to note that in recent times increasing numbers of people are beginning to accept this aspect of the paranormal, while other aspects are still regarded with much reservation. Thus we hear of military and police interest in the utilisation of so-called 'remote viewing,' but materialisation, dematerialisation, teleportation, etc., are still regarded as nonsense by many.

There are many well-documented and testified cases of a precognitive nature experienced by people the world over and of all walks of life, from housewives to physicists, lawyers, shopkeepers and children. One instance occurred in my own family, when a dear little girl of barely three years tugged desperately at her mother's skirts crying, "Look, Mummy, look, poor pussy hurt," pointing at an otherwise vacant spot on a grass verge, where only seconds *later* a cat lay dying, having been run over by a passing car and killed. This tragic event was witnessed by several horrified adults, who were naturally shaken by it. It is only one of many precognitive examples known to me personally.

I have noticed that the clarity of the event is often in direct proportion to its *emotive* content, while the more inquisitively inclined the recipient, the more detailed it is 'recorded.' Does that imply that very strong 'wishful thinking' can bring answers, which are otherwise before their time? In other words, is that very act a kind of *time tuning*? In my own case, I certainly can confirm the measure of intent, while a glance at the short list in the Appendix testifies to some of the results. Indeed, such mental exertions may have training qualities, for during my life I have on occasion had remarkably prophetic dreams which although of a purely mundane nature, the content was sufficiently unusual to register on the memory. The evidence therein has been accurate to a degree which would have been acceptable in a court of law. Therefore, certainly to my satisfaction, I have *experienced some future events*. Of course, one doesn't see every nut and bolt, nor every sentence with T's crossed, but imagery and basic functionalism is clear enough.

In a word, although I can neither prove nor fully understand this phenomenon, I am bound to accept that for myself or anyone else to witness future events, signifies that the brain of our species—in certain circumstances—can behave as a kind of time machine. In the following chapter we shall see a somewhat tentative explanation as to why this might be so. It also implies that we do live in a kind of 'eternal now' in which all events are prerecorded. In this respect, it is interesting to note this is entirely in accordance with the late J. W. Dunn's *Theory in an Experiment with Time* (1927) in which he described his own personal experiences which largely influenced his eventual theory.

Firstly, the term precognition can literally be translated as meaning *foreknowledge* of certain events. Secondly, Dunn was able to establish that not only *certain* people, but *everyone* dreams. Analysis of these dreams revealed many aspects of them are of a precognitive nature. It was primarily this weighty truth which led him to formulate his theory of space/time serialism. In his book he offered a technique whereby people who claimed not to have dreams can discover that they in fact *all* do. Among people who *know* they dream, it is quite common for them to realise that, although mixed with other meaningless paraphernalia, there is often a glimpse of an event which was yet to occur. Much of this is more often than not of a trivial nature, yet sometimes it is extremely difficult to deny or ignore. Like, for instance, this one which occurred to the author one night in the summer of 1950.

I had a very vivid dream in which I was kneeling beside a stretch of transparent water. My shirtsleeves were rolled up and my clenched fist was immersed in the cold water up to the wrist. Protruding from either side of my fist was a beautiful large goldfish, still alive, and in the dream I was squashing the poor creature to death! The dream fortunately terminated by my waking up in a cold sweat, thinking, "For God's sake why would I want to do such a thing?" It made a lasting impression on me and I put it down to a typical 'nightmare.' Then one *Saturday* afternoon—sometime after the dream—my wife Irene and I visited a property in the country which was approached over a small bridge spanning an attractive pond stocked with the largest goldfish I had ever seen. Peering approvingly down at them through the clear water, I was forcibly reminded of the dream, but promptly dismissed it as of little consequence. On the *following Saturday afternoon* we took our small son Gary in a rowboat on a park lake, in which there was a small island. As it was his first boat trip, the little chap had equipped himself with classic jam jar and net, just in case he could "catch some tiddlers."

Having reached the island, Gary became quite excited by the fact that several golden carp (silver in colour) were descaling themselves by threshing about near some small rocks and in the very act of lowering his small net, a carp leapt out of the water and nose-dived into it, literally it seemed, gave itself up! So comparatively large was the fish it promptly burst the net and ended up entangled with head out of the bottom and tail at the top. Our first reaction was to put it back into the water, but our little son was very disappointed; this after all was his very first 'tiddler.' So, in desperation, Irene held the fish's head in the water-filled jar while I rowed frantically to the quay side. Not really knowing what else we could do, she and Gary raced off to the boat shed and came back with a large empty paint tin, filled it with water and released the fish into it. But the fish began hurling itself out of the tin onto the ground as fast as we replaced it, not helped very much by a frantic child. By this time I noticed that the fish's gills were becoming caked with loose sand and having despatched Irene off to the boat shed to find a larger container, I did the first thing that came into my head; I grasped the poor thing and in *order to clean its gills* thrust it into the lake. It wasn't due only to the chill of the water that sent shivers down my spine!

Now here we have a dream, spread over, it seems, two separate weekends, both on a Saturday afternoon, both involving fish, one a *real* goldfish, the other a golden carp. One distraught son, two devastated parents, one of whom rather than h<u>urt</u> the poor creature as in the dream, was very anxious not to let it go on the one hand, but desperately anxious *not* to crush the life out of it on the other! Was there a happy ending? Well Irene *did* find a large saucepan, which with another net stretched over the top contained the fish until we arrived home, where it was transferred to a large tank for a while so Gary could show off his first *tiddler* before releasing it into a local stream.

Included in the Appendix are copies of published examples of the writer's conscious 'visions' which were apparently recorded at an earlier time to their eventful realisation, and therefore might qualify as being of a precognitive nature. Coincidentally, during the processing of these ideas, there were 'ordinary' precognitive dreams. In these cases there were details which included colour of materials, location of a model in downward out-of-control plunge, weather conditions at the time, etc., about which I couldn't have had prior knowledge. It must be understood this sort of phenomena *do* occur regularly, and the implications are profound!

Personally, I have always observed the same degree of unbiased assessment of such incidents as I have the more mechanistic problems. The point being, I urge the reader to bear in mind that much of the foregoing, presently labelled paranormal, will be more acceptable in the years after 2000. I ask forbearance if I suggest that the very kernel of this present book will prove to be no exception.

The proposition set out in the following chapter not only explains time travel as a possibility, but lays a foundation for suspecting the process might actually be a necessary one.

Telepathy

Of the various aspects of 'psychic' phenomena, at present telepathy is the most widely accepted. Most people at some time believe they have experienced it. Research under test conditions tends to support it, but at the end of the day investigative research scientists are seriously handicapped unless they have experienced it for themselves. The situation is synonymous with those scientists who debunk UFOs, until they have a 'close encounter,' and this author is acquainted with many of them. All the 'controlled' examinations in the world are of little use for true evaluation, for only personal experience will convince the otherwise well-meaning sceptical scientist that the subjects are neither conniving, lying nor deluded. The following are but two instances which happened to me which cannot persuasively be alluded to in the latter terms.

The first memorable occasion took place one weekend in 1944, after I had journeyed from Hatfield, Hertfordshire, to our home in Norfolk. At that time, I was engaged in the construction of the small turbojet engine discussed in Chapter 1. That particular week, I had completed a component which I brought with me for checking, so as to complete the job at Hatfield the following week.

By the time I arrived home, it was late and our small son Gary—then only three years old—was already tucked up and sound asleep, so after a meal and refreshment I was keen to check the turbine component. For this, I required a pair of drawing compasses from an old disused drawing instrument kit, somewhere in the spare room adjacent to Gary's bedroom. Having quietly negotiated the staircase in order not to disturb him, I opened the door only to discover that the light bulb had blown. Trying to imagine where the old instrument set might be, I carefully crossed the room and was in the process of feeling along a shelf when I thought I heard Gary saying something. "Damn," I thought, for I was sure I hadn't disturbed the little chap, so I went to his room, leaned over the bed and reassured him it was only me and was he OK. Again he murmured something as if half asleep, so I leaned closer and asked what he said. I stood in stunned silence when he repeated, "Are you looking for your compasses, Daddy?" Somewhat taken aback, I said I was and why did he ask, to which he drowsily replied, "I just thought." I looked carefully down at him; Gary was sound asleep!

There were a hundred and one things in the room that night, any one of which I could have been concentrating on in the darkness and that which I was trying to locate I hadn't used for ages. The following morning, taking care not to puzzle him, it became evident that Gary had no recollection of this event.

The second occasion took place one morning in the summer of 1961, when I was bent over my drawing board wrapt in concentration on a design layout. The background noise was moderate with the odd typewriter and general conversation, etc. Without conscious association of family matters, I suddenly received the thought, impression, call it what you will, of my mother in a distressed state emitting a shocking shrill scream. Puzzled and believing it was after all only a spurious though unaccountable errant thought, I was momentarily disturbed, glanced round the room a bit shaken and resumed my work.

Within half an hour, I received a telephone call from my sister informing me that my mother had indeed had an accident. On arriving at the hospital I discovered she had slipped on a garden step and badly injured the lower half of her leg, requiring over fifty stitches.

Later that day found me standing with my father at the spot where the accident had occurred. Through tearful eyes he looked at me and said, "Son, I shall never forget that dreadful scream." I had to restrain myself from answering, "Neither shall I, Dad."

Now this is a hard way to receive evidence of telepathy and as a meticulous investigator I must add, no sceptic in the world is going to convince me that I imagined these incidences. I *know* they occurred and the chief reason for including them here is to reinforce the argument that this is a natural personal phenomenon which has to be experienced. The alternative for the psychically barren sceptic is to at least try to accept that not all people everywhere are deluded or over-imaginative.

As I am inclined to the view that the phenomenon of telepathy—as with all other so-called paranormal phenomena—is an intrinsic facet of the overall cosmic matrix spectrum, it will be useful to consolidate this discussion with another illustrative example.

Let us assume that two people, A and B, are separated by a certain distance, near or far. In terms of the cosmic matrix, it follows that at any given time the material structure (atoms) forming the brains of A and B are constantly being fed by the creative rays which form them and, due to the omnidirectional nature of the rays, A and B are constantly in line. This is easier to understand if we neglect all the rays with the exception of those in one plane only, so that even if A were immobilised and B is orbiting around A, *always* the materials forming the brain structures are interconnected by their own private 'C' rays. As stated earlier, neither distance nor angular displacement makes one jot of difference. Therefore, once people *know* how to modulate (tune) the brain cells (atoms) and think of the other person (direction finding), the other person's brain cannot help but receive the signals. The *decoding* at the receiving end is, of course, naturally helpful! In terms of the Comat, this is what we call telepathy.

When you think about it, there is nothing new about this process, for when examined in terms of the Comat, it is precisely the system we employ every hour of every day. What difference there is, is one of degree not of kind. Viz., A thinks of speaking to B, who is five hundred miles away. Now the process becomes: find a phone booth. In transit to it, A's brain is never at any time disconnected from that phone booth, he or she never was, but now a conscious selection has been made. From then on, to the process of opening the door, placing the coinage, speaking to the operator or dialling the number, thinking of the person B, etc., the materials of all the structures taking part, no matter what movements or displacements there may have been, are *always* constantly interconnected by myriads of 'C' rays. In other words, the mechanical effect is much the same as that which we call telepathy, although this latter somewhat convoluted alternative process takes much more effort and time, and isn't as private as telepathy can be. *Perchance mobile phones may be a step nearer?*

Mental Bugging

Telepathy involves a kind of personalised bond 'tuning,' which—despite what some may be inclined to think—is incapable of being bugged, unless the sender and/ or receiver sanction it. For instance, I can impart certain information to an individual on the telephone, or I can choose to impart the same information to many individuals simultaneously at, say, a public gathering *with our joint compliance*, and telepathy functions similarly.

It should be understood the Theory of Unity logically identifies every act of thought energy as being a mini-transmitting station in its own right. So the sender doesn't have to be facing the direction of the other person involved for them to receive the signal. Thus—as with all radiant energy—only a small fraction of it is intercepted by the intended recipient, the main essence travels back towards S.

The Vibratory Logic for Sympathetic Thinking

During that journey, it will inevitably be vibrating in sympathy with other similarly motivated thought patterns. In the process, the combined signals can become cumulative and, due to the rotary nature of the creative rays, sooner or later, as sure as night follows day, they will return and be received by the original sender [Fig. 1.6].

Regretfully, the same is true for *all* radiant thought, not least the more negative variety. The extraordinary thing is, who told people in ancient times that there is a 'tenfold retribution' or self-inflicted 'Karmic' process? It was natural for those ancient ones to interpret this in terms of a Godly punishment—or blessing—as the case may be. Strangely, although we may now begin to understand this functionalism, in a way they were right. We can begin to realise there is a self-imposed penalty for attempted mental bugging, or telepathic interference, which is far more potent than the earthly legal kind. At the risk of a little over elaboration, how many of us would be quite so ready—for whatever reason—to hurl a stone, shoot an arrow or a bullet at an adversary if we knew for certain that the missile would travel in a circle and hit us on the back of the head? This is suitably analogous to the natural process indicated by the principle in this book. Once we can accept the logic of this basic theory, all kinds of imaginable situations and effects can be identified and reconciled—including those set out in the following chapter.

Clairaudience

Although there would seem to be a close affinity between telepathy and what is called clairaudience, there is a distinction. In the case of telepathy, the percipient receives an inaudible impression in which—as in the instance of my mother's accident—there can be a very defined though localised sound. Like the effect of wearing stereo headphones, the 'sound' seems to come from within.

Regretfully, very little is known about this phenomenon at present and as with any other human faculty—such as the arts—it can be controlled and used for benign purposes, or allowed to run riot and cause havoc in its wake. Such 'gifts' should be respected and used wisely. Of course this is not intended to imply that *all* such phenomena are intelligently engendered; similar effects can occur due to abnormality in the brain. In which case, often as not, it is healing the unfortunate percipient needs rather than psychological counselling.

Among those who have been fortunate to have been blessed with a more focused version of clairaudience are the spiritualist mediums, who claim to receive messages from the departed, and even UFO occupants. Interestingly, many of them have claimed to receive verifiable messages—in the audio sense—from the living, all of which seems to imply a common telepathic process is a root cause.

Psychometry

One cold winter evening in 1948, I had just settled myself alongside a welcoming fire in the house of a friend and his wife, Hilda, who had invited me there for a discussion on paranormal matters; in particular materialisation and levitation. Having made the acquaintance of Edward S. at my place of work, I learned that he was a particularly gifted physical medium, so I was fascinated at the prospect of witnessing a demonstration of levitation.

Edward's wife joined us in the living room with a warming cup of tea, but smilingly they said there was something they wanted to ask me before we began the evening session: If I were seated relaxing in a quiet darkened room, with or without eyes closed, they said, did I mentally see random shapes or even pictures? Somewhat taken aback, I affirmed that indeed this was so, but didn't everybody see the same thing? They were obviously pleased about this and after casting rather knowing glances at one another, they told me, no, not everybody does.

My puzzlement was further heightened when Edward produced from his jacket pocket a small box which he offered to me, asking if I would mind taking part in a somewhat impromptu experiment (though I couldn't help feeling there was something of prearrangement about this). He asked me to hold the box, close my eyes, take my time and describe anything I might see. I felt this was all quite amusing and agreed. Meanwhile Edward and his wife began moving the tea cups, etc., into the kitchen, talking quietly about this and that.

I was comfortably seated away from the chill of the night outside, relaxed, in good company, slightly amused, so I thought 'what the hell.' Settling back, I closed my eyes to look at the mental pictures as I have always done, and sure enough they were there, fleeting and ill-defined at first, so I responded to Hilda's request for an audible commentary.

In my mind's eye I was in the city of London, I was sure, although I couldn't identify the street. It was straight and flanked on either side by tall official-looking buildings. I was standing on the pavement below one of these buildings; people and traffic were everywhere. In the distance, some way ahead of me, my attention was drawn to the figure of a man walking towards me on the same pavement. As he got closer, my attention was drawn to the fact that he was wearing, of all things, a black top hat! He was tall, so that as he drew near my gaze was focused on his midriff, black unbuttoned tail coat exposing a grey striped waistcoat adorned by a very impressive watch chain.

Next I looked up to his face and saw that he sported an equally impressive greying military moustache and rather piercing eyes.

Still with my eyes closed, I knew that this impressive figure reminded me of someone, but who? Finally, my exasperating efforts to recall that someone were rewarded, when I realised this individual was none other than Neville Chamberlain (the British Prime Minister 1937-1940). Relieved at having put a name to my imaginary visitor, I then realised the chatter from the kitchen had ceased and looking up, I was greeted by Edward and Hilda standing in the dining room doorway with broad grins on their faces. Still smiling, Edward strode across the room, reclaiming the box he had given me to hold. As an inexperienced novice, I was bewildered and not a little centre-stage embarrassed as he related the sequel to this narrative.

Apparently, Hilda's father had been a porter in the Stock Market and in those days the city 'uniform' was indeed striped trousers, grey waistcoat, tail coat, cravat and top hat! Moreover they said that the old gentleman was so much like Neville Chamberlain, that he was chided by his peers in the city as "here comes the Prime Minister." Edward opened the lid of the box to reveal to me its contents: it proved to be a gold watch given to his father-in-law on his retirement, not long before he died. I had just received my first introductory acquaintance with the art of psychometry, which one day would take up so much of my time.

This account is, of course, quite true, but it pales into insignificance compared with other examples that have come my way. But as I have said it is the one I know best, which has helped to motivate me in trying to establish an underlying process of the kind I have offered in this book. It is anticipated that among those of my readers who accept some of these issues, the point will be raised that I may have simply been picking up the thoughts of my friends, in other words this may have been more to do with telepathy than psychometric response from the watch. However, I have other reasons for believing this wasn't so in this instance. It is well known to students of this subject that efficient mediums will often reveal verifiable information they have discerned from long-buried objects, etc.

Remote Viewing

This is another similar, mentally discerned phenomenon commonly associated with the OBE (out-of-body experience) in which the percipient is able to view sometimes faraway places and people. Regretfully, it is symbionic of the times that individuals possessing a highly developed degree of this mental ability are sometimes called upon to act as—what in the final analysis is—little more than psychic sniffer dogs against crime; whereas it would be reassuringly nice to hear of this gift helping to save terrified victims of earthquakes around the world, for instance, which is probably more in line with the *original* design among living creatures.

Apparitions and Haunted Places

Over the years, my enquiries have taken me into the realms of mysticism, spiritualism, healing and paranormal phenomena. I have both seen and spoken with allegedly departed spirits of the so-called dead. I have met with and listened to witnesses—

some among my own family—who have seen apparitions of people from an earlier time. One aspect which is common is the fact that the percipient is unaware that they are seeing something out of the ordinary and despite all my acquaintance and familiarity with the subject, when I saw an apparition I was no exception to this reaction.

Sometime in 1978, I took my Labrador dog Jason for his usual late evening walk, which took us down to the nearby beach. This particular spot bordering the Solent was both pebbled and sandy with the occasional sloping stone-built groyne, over which we had to clamber. One of these terminated at the shoreward end in a pile of rubble, being the remains of a wartime pill box. As this was a favourite spot for local anglers, much to my embarrassment Jason had been known to scrounge some of their sandwiches and even bait! So I was usually relieved when the tide was out and we had the beach to ourselves—as we did that evening. On this occasion I was approaching the first stone groyne, which Jason had already taken in his stride, when I suddenly noticed someone coming towards me carrying a lighted lantern. Calling Jason and hoping he wouldn't make a nuisance of himself, I prepared to meet this individual as we approached the groyne from opposite sides.

So far there had been neither a response from Jason nor a greeting from the lantern carrier as we both reached the groyne. I jumped over, expecting this person to do the same, and was surprised to find neither lighted lantern nor bearer there! Puzzled, I looked around in the gloom. There was nothing, only boulders and beach debris to be seen. I looked up and down the groyne, reasoning as I did so that it was impossible for anyone to have reached the tree lined shore *noiselessly* in seconds. Yet somehow, someone had.

Jason and I resumed our walk, and I dismissed the matter until the following evening, when I realised that all nighttime anglers used very brightly-lit gas lamps, whereas I had seen a much dimmer light of the kind emitted from old-fashioned oil lanterns. The type, in fact, that anglers of earlier times would have used. Extraordinary as it may seem, that particular part of the beach had a reputation for such strange happenings, but for years I had taken this walk with no such thoughts on my mind.

At the risk of straying a little out of context, that particular part of my preamble has other memories for me. Again, while walking with Jason one early evening before dusk, I had eased my stroll to a standstill, scanning the skies for an oncoming low-flying aircraft. Despite the fading light, the sky was clear and any moment the aircraft should have swept into view over the Solent. Instantly, I detected the—to me—familiar[2] heterodyne beat of Merlin engines as fitted to the Lancaster and Halifax bombers during World War II.

With ever-increasing nearness and rising roar, I thrilled in anticipation of such an unexpected reunion with one of those—now rare—flying machines. The noise bore down on me, sweeping past with that unforgettable throb, throb. By now I should have seen it in all its memorable glory. But it swept onwards and upwards into the clean evening air, yet I saw absolutely nothing! Transfixed, I stood there staring unbelievably as the ever-diminishing engine note finally faded.

Extremely puzzled and not a little disappointed, I began to resume my stroll with Jason. Then there it was again; once more approaching from the rear. The aircraft had made a complete circuit and was doing another run. This time I wouldn't miss it, yet I did; and the same thing all over again. Three unbelievable circuits before my very eyes, yet I hadn't seen a thing.

Astonishingly, although I had long since dismissed the matter, over the next two years I was treated to no less than three repeated aerial performances—at the same spot—on the last occasion right over my head, but this time as a measure of reassurance there were other people present who heard the same thing. Logically, this item should have been included under the heading of clairaudience, but its association with the *place* of origin perhaps appropriately warrants it. This particular aspect is more fully discussed in the next chapter.

Exiles from Time?

As stated, there are many, many examples of hitherto puzzling aspects of nature which I have no difficulty in relating to the Theory of Unity, and of necessity those that have been covered here have been scantily addressed. The following example is no less important—indeed some may say it is *the* most important of them all—yet regretfully our available space here deems it to be no exception. Despite the inherent risk in this respect, it is hoped due allowance will be made for brevity.

Due to my involvement with the matters forming the structure of this book, it is inevitable that I should find a persuasive coherence among it all, with the Theory of Unity, the cosmos, and gravity at one end of a broad spectrum and aerospace and paranormal activity at the other. Of course, it is accepted many will disagree, yet equally I have known many who, like myself, have never had difficulty in logically correlating it all.

Following the proposition that the source in which our universe has its being, not only is the origin of the space-forming rays in which all matter is manifest—including the multi-duplicate parallel forms—also creates another kind of dimensionless space in which the pure mind stuff of all living creatures is primarily manifest. This place, this domain of pure light and unadulterated energy is sometimes commonly called Nirvana, Heaven or the Spirit World, etc., depending on one's religion. I venture to suggest that everyone sometime, somewhere will eventually come to acknowledge the veracity of this truth.[3]

If I might be indulged an even more philosophical posture, I further suggest that in the final analysis it will be discovered that the planet Earth is a remarkable—in cosmological terms—expensive and rare learning habitat; in other words, it is a school for embryonic intelligences in the making. Is it a coincidence that through the ages prophets have continued to assure the rest of us? For instance, Christ is quoted as saying, "Oh ye of little faith, know ye not ye are all God's and the works I do now so shall ye do, and even more wondrous works than these shall ye do." He is also quoted as saying, "Had ye faith the size of a mustard seed ye could remove mountains." If a lonely man said that 2,000 years ago, how did he and other prophets know the truth of these matters?

The pure mind stuff, or soul if you wish, craves and is sustained by experience—of all kinds—just as surely as the body craves for and is sustained by bread. So a consensus would be that whether we like it or not, as voluntary exiles *we* chose to leave that beautiful state to come here as part of a learning course. On arrival, the die is cast and *everything* which we *chose* to happen *will* happen. From the time of birth to the time of death the process might vary a little along the way, but the main course will be followed to fruition. This is the only reconcilable truth which indicates that there *is* such a thing as 'free will' and there is such a thing as fate, or destiny as the phenomenon of precognition, for instance, forcefully illustrates. It is the only fair answer to the sadly oft-repeated question, "To allow all this suffering can there really be a God?"

This school, this place of cosmic learning, is subject to all the rules of the material universe where atoms, quantum physics, stars, galaxies and physical life of all kinds endure. Our Earth is but one of the 'many mansions' and the human race is merely one of the 'other sheep' Christ said he 'had to bring.' Moreover, I suggest that it will be discovered that all matter, from the microcosm to the macrocosmic scale, is imbued with a degree of autonomy or self-regulation within the Cosmic Matrix. In other words a planet, far from being composed of inanimate matter, is in fact a complex living organism, with a state of intelligence far removed from our present understanding. In which case we shouldn't be surprised to learn that it is capable of relating to the collective mass thought patterns of mankind—together with all that that implies.

Summarising this first part of the paranormal attributes of the Cosmic Matrix indicates that although the basic effects seem to be more of a mental rather than physical nature, we should remember that even this process has its roots in the wave formations of all the so-called physical components involved. Any apparent difference between these effects are exactly that—apparent—as illustrated in the next chapter.

[1] The brilliant mathematical physicist and author of *Black Holes* and many other works.

[2] Heterodyne effect, caused by superimposing two identical slightly out of phase frequencies to produce a third; in this case a sonic beat.

[3] Einstein is known to have held similar views.

7

PARANORMAL ATTRIBUTES
Of the "Nuts and Bolts" Kind

This second half of the apparently diverse list has been selected not so much due to the fact that they *are* different, but that the *effects* in other wave structures in the Comat are more definable, in short they are more physically noticeable. So, whereas the previous apparently mental phenomena—in a high percentage of cases—are experienced only by the percipient, in the more physical kind the process directly involves a change of state at the atomic level, hence the tongue-in-cheek title of this chapter. However, as in the last chapter, throughout the following I shall try to remind the reader now and again that in both the mental and physical attributes, in essence we are merely dealing with wave mechanics.

As with the previous cases, there are subdivisions of the following, but the list contains those which are broadly appropriate to the main emphasis of the book.

1. Levitation
2. Radiesthesia (divining of objects and water)
3. Teleportation (dematerialisation, materialisation, dimensional change, parallel universes and time travel)
4. Invisibility
5. Electric people (and effects on machinery)

As a result of my acquaintance with so-called paranormal effects, I have formed the opinion that a natural development of the Unity of Creation theory suggests that all living creatures have a bioplasmic extension to the existing auric field. This is more pronounced in the human species, some more than others.

In certain circumstances this bioplasmic extension can subconsciously, or consciously, serve as a triggering or 'tuning' mechanism, which for an immeasurably small amount of initial electrical input can unleash comparatively vast amounts of Comat energy. It is commonly termed psychokinetic or PK energy. It is this which can be responsible for certain dramatic effects, such as

*A more passive demonstration
of levitation with the world
famous medium Jack Webber
under test conditions*

*Fig 1.7. Terrified Madame
Costa clutches her baby
during the onslaught of
poltergeist activity*

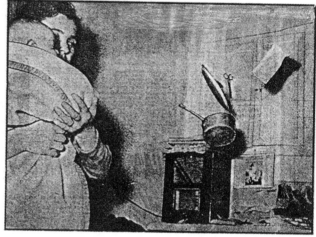

seemingly miraculous levitations.

Paranormal Levitation

Of the many paranormal effects, levitation remains not only the most impressive, but it is crucial to the main purpose of this account. Certainly there exist far more quantitative sources of information, but the following cases have been included due to their technically useful content, which will become apparent to the discerning reader.

Special Report by Radio Sottens, Studio de Geneve, March 7th, 1955
Monsieur Escoffier, appointed reporter of the Geneva Studio, tells the following story in a voice which still echoes his strong emotion and perplexity:

150 km south of Geneva, in Savoie, France, lies the village of St-Jean et St-Julien de Maurienne. Three months ago, the wife of a poor Italian mason moved into this community to join her husband who works temporarily there. Two months ago Madame Costa had a baby. The family Costa lives in a very humble old dwelling in rather poor circumstances.

Ten days ago, around February 25th, 1955, a phenomena started: pans, lids and furniture started flying and moving around without being touched by anybody. This happened only when Madame Costa was near. The kitchen table went up, the kitchen stool moved sidewise on the floor; the cupboard uttered strange noises and there seemed to be heavy beating inside. A water glass fell from the top without breaking, another one remained on the top but broke in two. All these things made strange noises when moving through the air.

When Radio Sottens heard of this Monsieur Escoffier, a reliable reporter for scientific news was sent to investigate. He found the Costa house full of frightened people who had witnessed many of these incidents. A very strange story was about a hot water bottle (bottle-shaped, earthenware) safely tucked away in bed which left the bed by itself, moved into the air producing the same strange noise and fell to the ground with a very strong crack. All witnesses testified to this noise being much stronger than an earthen object would be expected to produce when falling to the ground. The bottle never broke, and when lying on the ground it never rolled over. It just **stayed** where it seemed to have been put. And this strange phenomenon happened again when Monsieur Escoffier was present! One of the strangest facts to him seemed that the bed was not undone when the bottle left it.

Once, he was told, Madame Costa sat on a bench when she together with this bench was lifted into the air where she remained some moments. The bench then fell heavily to the floor, she herself was put down rather gently without being hurt. Another time, a young man tried to catch the bottle in the air, but when he touched it *Madame Costa fell to the floor* and remained unconscious for 50 minutes.[1]   Therefore, nobody dared to do this again. The poor woman is frightened out of her wits, she prays all day long to all saints she knows, holding her rosary and a cross between stiff fingers. She could not be interviewed as she talks no French.

A priest who was called said that this looked like a case of the Devil! And that if things did not stop he would have to ask for a special "exorcising priest" from Rome!

There were at least 50 people around who had witnessed and were ready to testify. [See Fig. 1.7]

In 1977, in a 40-minute BBC Radio 4 programme, listeners heard tape-recorded supernormal rappings. Two independent investigators, one a national paper's photographer, the other an engineer, testified that a poltergeist was responsible for 'impossible' happenings to their equipment. Scientists, the well-known psychical researcher Maurice Grosse, psychologists and other experts all failed to pinpoint the source of the paranormal happenings which occurred at a Middlesex council house in Enfield. On one particular occasion there were no less than eleven people present when things started flying around. Children's bricks were thrown, and though nobody moved an arm, one brick struck the photographer, giving him a lump on the forehead which lasted for four days. In all, the phenomena (which included all kinds of supernormal effects) included the levitation of a child, Janet Hodgson. On one occasion, a school 'lollipop lady,' Hazel Short, was on duty near the house. Returning at lunchtime, she spotted a red cushion on the roof. Mrs. Short asked one of the family's daughters how it got there, but she had no idea. She said, as they stood talking and watching it:

I suddenly heard a bang. I saw a book hit the front bedroom window. That was followed by a pillow. Then there was another book and the pillow again. I saw Janet going up and down the front of the window. I thought she was jumping, but when I looked she was horizontal. Her arms and legs were going everywhere.

John Rainbow, a baker's roundsman, also witnessed the girl's levitation:

Before that day I would never have believed anything about it. The child appeared to float half round the room. At the same time the curtains were blowing into the room, though the windows were completely closed. Something inside seemed to be drawing the curtains off the window. The articles and the child seemed to be revolving around the room clockwise. The child's arm banged against the window frame twice. I was frightened, because of the force with which she banged against it, that the frame would go. I fully expected her to drop into the road.

Ron Denny, the BBC engineer, said that his video equipment was adversely affected; it was hoped to obtain photos with automatic cameras. He said:

On one occasion I went through the start sequence, pressed the button and all the lights on the recorder came on one after the other. This is absolutely impossible. There is no way we know to make this happen. The recorder had particular facilities for editing, sound dubbing and so on. Each had separate buttons you had to press. When you do this they light, there was no logical way they could have lit simply by pressing the 'on' switch.

The machine also jammed completely. When we eventually

managed to retrieve the cassette from the machine we found the tape had come out of the cassette, it had wound itself round part of the machine. We have never had this happen before. It is probably one chance in a million.

Here is an account by Sir William Crookes (1832-1919) who witnessed several levitations by the well-known medium of that time, Daniel Douglas Home.

The best cases of Home's levitation I witnessed were in my own house. On one occasion he went to a clear part of the room, and, after standing quietly for a minute I saw him slowly rise up in a continuous gliding movement, and remain about six inches off the ground for several seconds, when he slowly descended. On this occasion no one moved from their places. On another occasion I was invited to come to him, when he rose eighteen inches off the ground, and I passed my hands under his feet, round him and over his head when he was in the air. On several occasions, Home and the chair on which he was sitting at the table rose off the ground. This was generally done very deliberately and Home sometimes then tucked up his feet on the seat of the chair and held up his hands in full view of all of us. On such occasions, I have gone down and seen and felt all four legs were off the ground at the same time, Home's feet being on the chair. Less frequently the levitating power extended to those next to him. Once my wife was thus raised off the ground in her chair.

The frequently reported 'miracle' of Home's body floating out of windows over 60 feet above street level occurred in London on 13th December 1868, in the presence of three witnesses. These were Lord Adare (later Lord Dunraven), his cousin Captain C. Wynne, and Lord Lindsay (later Earl of Crawford and Balcarres), who at one time was a member of the Council of the Royal Society.

Home's extraordinary ability to be able to levitate is by no means unique. In the East, levitation of the human body is an accepted fact. The Hindus claim that they are able to produce the phenomenon. They stress the importance of a *proper mental attitude* together with certain physical exercises, including deep rhythmical breathing.

It is claimed by certain adepts that by closely observing the prescribed ritual, a "living force can be generated which counteracts the force of gravity."

Over the years there is continual reference to people experiencing autolevitation. The French Abbot Augustine Calmet wrote in 1751:

Persons full of religion and piety who, in the fervour of their orisons, have been taken up into the air and remained there for some time. We have known a good monk who rises sometimes from the ground and remains suspended without wishing it, without seeking to do so, especially on seeing some devotional image, or on hearing some devout prayer such as Gloria in Excelsis Deo. I knew a nun to whom it has often happened in spite of herself, to see herself thus raised in the air to a certain distance from the earth; it was neither from choice nor from any wish to distinguish herself, since she was truly confused by it.

Calmet also related the story of St. Richard, Abbot of S. Vanne de Vordum, who in 1036 was "elevated from the ground, while he was singing mass in the presence of Duke Galizon, his sons, and a great number of lords and soldiers." And another later levitation, that of Father Dominic Carmo Dechaux (Dominic of Jesus-Mary) in Madrid in 1601, in the presence of King Philip of Spain, his Queen and the entire court. "So that they had duly to *blow upon his body to move it about like a soap bubble*." Doesn't this suggest an absence of inertia, or a kind of gravitational 'transparency'? Before some readers raise their eyebrows and purse their lips in doubt, may I respectfully remind them that they probably witnessed demonstrations of field effect 'transparency' in their early school years, only then they were totally unaware of the real cause. For the benefit of the unacquainted, I refer to the simple experiment involving magnetism in which a piece of iron is held suspended by a magnet. If the iron is heated, the atoms contained in it quickly reorientate from alignment to their original random state, so that, in conventional terms, the magnet becomes ineffectual over the iron, causing it to fall. Now if, as I suspect, magnetism is akin to gravity in terms of the Theory of Unity and bearing in mind that not all the atoms in a magnet are unbalanced, then indeed we can justifiably describe the foregoing phenomenon as a kind of magnetic 'transparency.' In which case, again in terms of the Theory of Unity, we shouldn't be surprised if a state of *gravitational* transparency does sometimes occur.

Perhaps one of the most complete works of autolevitation is the book of Olivier Leroy, the French university professor, entitled *Levitations*. In this book Leroy repeats some two hundred instances of levitation. Among them the story of St. Teresa of Avila (1515-1582), the famous nun who was responsible for the drastic reforms in the Carmelite Order in Spain. She was canonised by Pope Gregory XV in 1622. Here is a somewhat condensed account of her own reactions to her unusual experience.

> During rapture, the soul does not seem to animate the body . . . rapture, for the most part, is irresistible. It comes, in general, as a shock, quick and sharp, before you can collect your thoughts or help yourself in any way, and you see or feel it as a cloud or a strong eagle rising upwards and carrying you away on its wings . . . I would very often resist and exert all my strength, particularly at these times when rapture was coming upon me in public . . . Occasionally I was able, by great efforts, to make a slight resistance; but afterwards I was worn out, like a person who has been contending with a strong giant; at other times it was impossible to resist at all; my soul was carried away, and almost always my head with it, and now and then the whole body as well, so that it was lifted up from the ground . . . It seemed to me, when I tried to make some resistance, as if a *great force beneath my feet lifted me up*. I know nothing with which to compare it . . . for it is a great struggle, and of little use, whenever our Lord so wills it . . .
> I confess that it threw me into a great fear, very great indeed at first; for when I saw my body lifted up from the earth, how could I help it? Though the spirit draws it upwards after itself, and that with great sweetness, if unresisted, the senses are not lost; at least I was so much myself as to be able to see that I was being lifted up . . . When the rapture was over, my body seemed frequently to be *buoyant, as if all weight had departed from it*; so much that now and then I scarcely knew that my feet touched the ground.

Of no less significance are the recorded stories concerning the monk St. Joseph of Copertino (1603-1663). Not long after his admission to the priesthood in 1628 and from then on until his death, Joseph experienced a large number of these levitations during which he sometimes enjoyed ecstatic flights when his body was transported from one place to another. It is said that for thirty-five years his superiors were compelled to restrict him to certain processions, even to the extent of some common meals in the refectory, because of the disturbances his frequent levitations would cause [Fig. 2.7].

Of the testimonies available, probably the most reliable is that of a surgeon, Francesco Pierpaoli, who says:

> During the last illness of Father Joseph, I had to cauterise his right leg by order of Doctor Giacinto Carosi. Father Joseph was sitting in a chair, with his leg laid on my knee. I had already begun

cauterising, when I noticed that Father Joseph was rapt out of his senses; his arms were outspread, his eyes open and lifted to heaven. His mouth was wide open, his breathing had nearly stopped. I noticed that he was raised about a palm over the said chair, in the same position as before the rapture. I tried to lower his leg down, but I could not; it remained stretched . . . In order to observe Father Joseph better, I knelt down. The above-mentioned doctor was examining him with me. Both of us ascertained undoubtedly that Father Joseph was wrapped in ecstasy and actually suspended in mid-air as I have already said. He had been a quarter of an hour in this situation, when Father Silvestro Evangelista, of the monastery of Osimo, came up. He observed the phenomena for some time, and

Fig 2.7. A drawing by G.Cades of one of St.Joseph of Copertino's ecstatic flights.

commanded Joseph under obedience to come to himself, and called him by name. Joseph then smiled and recovered his senses.

The 'Delights' of Levitation

Naturally, people are inclined to attach little or no significance to such accounts which occurred so long ago on the grounds that (granted special circumstances) if such phenomena did spontaneously occur, then surely it would be more widely known; and in any case how much reliance can be placed on such old records? In fact, due to the very public and ecclesiastical nature of these, quite a lot. Hoax is quite out of the question for such perpetrators would have required a technological backup that even David Copperfield would envy. As stated, today levitation still occurs among devout mystics and many genuine materialisation mediums who find it more prudent to keep a low public profile.

It seems to me that in the 16th century St. Teresa and Fr. Dominic Carmo and others like them were doing their best to describe a sensation which was alien to them, sometimes exciting, sometimes terrifying, but always different, but the fact remains that in the 20th century *astronauts are describing identical sensations!* Today we call it 'weightlessness.' One thing they have in common, they are all describing *a changed state.* Once more the question has to be addressed, is this *another* coincidence?

Between the years 1870-1873, Sir William Crookes conducted many experiments into levitation; the following is a quotation from an article of his published in the *Quarterly Journal of Science*, January 1874:

On five separate occasions a heavy dining table rose between a few inches and one and a half feet off the floor, under special circumstances which rendered trickery impossible. On another occasion a heavy table rose from the floor in full light, while I was holding the medium's hands and feet. On another occasion the table rose from the floor, not only when no person was touching it, but under conditions that I had pre-arranged so as to assure unquestionable proof of the fact . . . On one occasion I witnessed a chair, with a lady sitting on it, rise several inches from the ground. On another occasion, to avoid the suspicion of this being in some way performed by herself, the lady knelt on the chair in such a manner that its four feet were visible to us. It then rose about three inches, remained suspended for about ten seconds, and then slowly descended.

Upton Sinclair, author of the interesting book *Mental Radio,* which although concerned mainly with telepathy, has a very interesting description of a levitation by a young foreign medium, that he witnessed. Sinclair was never able to detect any trickery and he describes the amazing phenomenon in the following words:

> In our home (the 'psychic') gave what appeared to be a demonstration of levitation without contact. I do not say that it really was levitation; I merely say that our friends who witnessed it—physicians, scientists, writers and their wives, fourteen persons in all—were unable to even suggest a normal method by which the event could have happened. There was no one present who could have been a confederate, and the psychic had been searched for apparatus; it was in our home where he had no opportunity whatever for preparation. His wrists and ankles were firmly held by persons whom I knew well—and there was sufficient light in the room so that I could see the outline of his figure slumped in a chair. Under these circumstances a thirty-four-pound table rose four feet into the air and moved slowly a distance of eight feet over my head.

From the book, *The Unknown Effects of Mind on Matter* by Doctor Eugene Osty and Marcel Osty, comes more evidence of levitation. The investigators conducted many meticulously prepared experiments with the Austrian medium, Rudi Schneider, brother of the better-known Willy Schneider.

They did not seek sensational results, but spent a great deal of time in devising delicate and ingenious instruments in order to detect the psychic emanations from the medium.

Their findings convinced them that Rudi Schneider could mentally produce the formation of an 'invisible substance.' From time to time he would announce exactly when the force would affect the apparatus and this was always confirmed by the instruments. These investigators substantiated the claim that a *white light completely precluded the phenomenon,* while even a too strong red light interfered with it. (Note the significant connection between *this type of phenomenon and light.*) They used instruments that recorded unusual infrared ray disturbances and discovered that even these rays seemed capable of destroying the emanations.

There was the occasion when by sheer effort of mind, Rudi caused a flower to be raised from a vase and set it sailing over the heads of those present.

There seems little doubt that Alfred Still, author of the book *Borderlands of Science*, comes very near to the truth when he says:

> Granting that levitation does actually occur, it would seem that the mind of a living organism is capable of creating something physical—the psychic force, or 'invisible substance' of the Ostys, and that darkness, or the *absence of disturbing light*, is desirable while the act is pending. The physicist, who knows nothing about the nature of gravitation and very little about the nature of light, is inclined to believe in a streak of similarity between the two. This suggests that, just as light appears to weaken or destroy the psychic force that produces levitation, so this psychic force, in turn, may be able not merely to annul but *to control gravitation*.

The great physicist, Sir Oliver Lodge, once said:

> Life and mind and consciousness do not belong to the material region. Whatever they are in themselves, they are manifestly something quite distinct from matter and energy, and yet they utilise the material and dominate it . . . Mind does not itself exert force, nor does it enter into the scheme of physics, yet it indirectly brings results which otherwise would not have happened . . . A bird grows a feather, and a bird builds a nest; I doubt if there is less design in the one case than in the other. How life achieves the guidance, how even it accomplishes the movements, is a mystery, but that it does accomplish them is a commonplace of observation. From a motion of a finger to the construction of an aeroplane, there is but a succession of steps. From the growth of a weed to the flight of an eagle—from a yeast granule at one end, to the human body at the other—the organising power of life over matter is conspicuous.

In this context, may I again remind the reader of the sailboat and little boy on the escalator analogies we saw earlier, that of a comparatively small force acting as an agent to harness a considerably larger one. Well, the situation is exactly analogous in the paranormal instance: The brain is an organ composed of atoms ('C' ray modulations); as we have seen in the process of thought these radiate more modulations on the carrier wave (thought waves). At a certain frequency, these can act as a 'catalyst' to unbalance the extension modulations of the atoms which compose the structure of a normally heavy object. It is moved, not by the original infinitely weak thought waves themselves (or psi), but by a comparatively small proportion of the colossal power inherent in the Space Matrix. The real mystery is, how does the operator's brain know how to relay such a process? This is not so strange, however, when we consider that from beginning to end it is a function of a coordinated series in the Matrix.

In a word, so-called paranormal levitation, far from being an unbelievable phenomenon, could be the norm, that is, when we know how, people could fly—without wings. Thus "Superman" may not be so far out after all! The fact that so few of us can levitate is only a measure of the relative developed status of the rest of us and it may be significant to note that it is usually spiritually evolved people or so-called mystics who can. Put in more down to earth mechanistic language, they may have an innate ability to control or *finely tune* their 'C' wave modulations—necessary with any kind of sensitive electronic equipment—while the rest of us can't.

Even so, let us take heart, for we *can* fly (albeit *with* wings) and as we shall see, we will be able to traverse space in gravitational spaceships. Perhaps even before that, we may enjoy transport employing other means of propulsion in vehicles by which standards our present concepts will one day be regarded as primitive.

As stated, given the amassed material available elsewhere, a quantitative review of paranormal phenomena is unnecessary here whereas a qualitative approach may be more useful. For instance, notice that *emotionally* induced trance levitation is of a *passive* nature, i.e., there appears to be a total loss of inertia with no surrounding external disturbances, neither atmospheric nor involving locally situated people. Whereas poltergeist-type induced levitation is of a more *dynamic* and extensive nature, in that it *does* cause atmospheric effects and displacement of locally situated objects sometimes *in a circular motion*. In terms of that which has been discussed earlier, this would suggest that in the case of Father Dominic Carmo Dechaux and St. Teresa, these are not so much cases of an 'anti-gravitic' nature; rather they appear to suggest molecular, or atomic, 'transparency.' The more dynamic instances suggest concentrated secondary forces at work, more akin to 'anti-gravity,' of the kind which can produce a thrust reaction in any direction, rather than a kind of benign spatial isolation.

Radiesthesia and Dowsing Laid Low

No doubt one of the oldest techniques for tapping the Comat, water dowsing is but one aspect of a broad spectrum of paranormal phenomena known as Radiesthesia, which has become almost 'respectable.'

For instance, people have now begun to accept, though not understand, that so-called divining of water or other substances is a fact, while even large oil companies are known to employ dowsers very profitably.[2] In order to explain the modus operandi of this technique, it is claimed that the diviner relates to certain underground currents, etc. But what nobody has so far been able to explain is how the dowser is able to make contact with these currents via a map of the location, which may be situated perhaps hundreds or thousands of miles away!

I suggest that much of the mystery surrounding this is reduced by an acceptance and understanding of the fundamental principle behind the Theory of Unity. For once two particles (modulations) are formed, no matter what distances or angular displacements occur between them, they can *never* be separated from the matrix. Thus, there are *always* rays interconnecting them until they cease to exist in modulation form, and no matter by what intervening processes the map was produced; from the time measurements were first recorded on the earth by an unknown engineer, to the time when such information was merely thoughts in his mind, *never* has there once been separation of the interconnecting space modulations (atoms) which for ever afterwards would be sympathetically recorded in space for some future diviner to attain *tuning*. It is probably this process that pigeons, dogs and cats use to find their way home, as no doubt humans did at one time. It is the process which is involved with *all* radiant phenomena, be it light, radio waves, telepathy—and whether some of us like it or not—healing and every kind of so-called magic, including the incredible power of people like Uri Geller!

Teleportation (Mat/Demat)

The term teleportation was first coined by the prolific writer of unusual phenomena Charles Fort (1874-1932) whose books were collected into one volume titled *The Complete Books of Charles Fort* (1974).

One of the earliest recorded accounts of human teleportation—brought to light by the pioneer Ufologist M.K. Jessup—was recovered from Spanish legal records concerning the trial by the Inquisition of a soldier who, on the 25th October 1593, suddenly appeared in a confused state in the main square of Mexico City, wearing the uniform of a regiment stationed *nine thousand miles* away in the Philippines! The poor man could only say that moments before his appearance in Mexico he had been on sentry duty at the Governor's palace in Manila, the capital of the Philippines. He said the Governor had just been assassinated and he had no idea how he came to be in Mexico.

Months later, a ship arrived from the Philippines, and the news of the assassination together with other details confirmed the soldier's story—regretfully it seems—too late to do the poor fellow any good. There are other cases like this, some of which have appeared in verifiable circumstances in more recent times. This information is available to anyone who would take the trouble to seek it.

It is accepted any process involving teleportation must presumably include the dematerialisation and materialisation of matter (more colloquially known as Demat/Mat). Naturally, living in an illusionary world of 'dense matter' as we appear to do, it is extremely difficult to visualise such a process. But even the somewhat crude physical analogy offered in Chapter 3, where the two transient illusionary globes can be made to interpenetrate and pass through one another, may help in taking the first step towards this. Brought to its logical conclusion, it might be said there is so much surplus space *within* matter and just as much *outside* it, that *if* we could synchronously harmonise the vibratory motions of the particles (waves) forming our bodies with those of a brick wall, for instance, we could pass through it in a similar manner to that described in the case of the illusionary globes.

Teleportation is not a phenomenon most of us are likely to experience in a lifetime, yet as paranormal events go it is not uncommon, occurring all over the world every year. In fact, it used to be a constant source of amazement to me that rather than accept such extraordinary phenomena as literal research gifts from heaven—in the strictly physical sense—many scientists not only have the gall to deny it, but shun it like the plague! Here is the first of three incidents occurring in the south of England, which I investigated to my absolute satisfaction.

It occurred just after 12 a.m. on a warm summer night in 1961 in a house over which a low altitude UFO incident had occurred, which although observed by other parties from a quarter of a mile away, was *not* observed by the occupants of the house tself.

Visiting the designated area the following day I located the property—a large secluded Georgian house—and enquired of the occupants if I might ask them some questions. On being cordially invited in, I ascertained that the owner of the house (Mr. A) occupied the ground and first floor while two elderly ladies rented the second floor apartment. Mr. A was completely unaware of any external incident, but he was rather puzzled about my enquiry, for it became evident that he did in fact have a *very* disturbed night.

It transpired that just after midnight someone—he thought—had not only trespassed into his garden, but for some unaccountable reason hurled a large stone through the open window, which crashed on the floor and rolled across the room, awakening him. Thinking he had a drunk in the garden, he had picked up the stone and gone downstairs to investigate. On going out into the garden, Mr. A was annoyed to find his dustbin lid removed and hurled some considerable distance from the dustbin, while the contents were left untouched. As he had suspected a tramp or even a fox might do this, he was additionally puzzled and returned back to bed.

Mr. A was very accommodating and took me over the house and later the garden. The first thing I noticed was that the large stone seemed to be one of several Mr. A used as an ornamental garden edge. It measured approximately 5.5 x 4.5 inches and weighed several pounds.

The sash window in Mr. A's bedroom had been open by a foot or so and outside this was girdled by a small window balcony with an 18-inch-high masonry wall. Below the window was a stepped terrace, while the garden sloped fairly steeply away from the house. The height from the most suitable part of the terrace to the window was about 17 feet. Now I know of no athletic cricketers who could have—in the darkness—had the strength and skill to hurl that stone with the required force, and trajectory, up and over the balcony wall and neatly into the open window. I was sure it was impossible and said as much. Of course, Mr. A had already reluctantly arrived at the same conclusion, with the proviso that the person intended to smash the window. But this still didn't explain *how*, for it would have required a world shot putter par excellence to have achieved it.

And there the matter would have ended but for the fact that a few weeks later, a friend of mine happened to mention that he knew the two ladies who occupied the floor *above* Mr. A, and visiting there he had out of interest asked them if anything unusual had occurred that particular week. They had looked at each other rather shaken and somewhat diffidently confided in him because they "didn't want to offend Mr. A." Apparently, that very same night and at the same time, *two* similar stones were dropped into *their* bedroom with a thunderous crash and frightened the life out of them.

Now it seems we had *some* athlete on our hands, for the height from the terrace to the ladies' bedroom window was no less than 27 feet. But that became rather academic when I learned that the two dear souls *never* slept with the windows open and their bedroom door was *always* locked! Later, Mr. A replaced the boundary stones in their located places, leaving one bewildered UFO investigator to ponder, not so much on whether this thing occurred, nor even how, but Why? The beginning of an answer to that lies somewhere between this chapter and the included Appendix.

Such "apports"[3] are by no means rare; falls of stones, coins, brushes and all kinds of artifacts have materialised between floor and ceilings, a phenomenon which sometimes accompanies poltergeist activity, but not always, as was the next case which also involved two people well known to me.

A Worthless Trinket?

For many years Tom and Win were very dear friends of ours. At one time Tom had been a Co-operative Store manager, but earlier on in life he had discovered that he had a natural ability of contact healing (laying on of hands). Being one of those good souls who couldn't refuse a plea for help, Tom's reputation as a healer soon got around, particularly when it became known that one of his successfully treated patients was the then Vicar of St. Albans, Hertfordshire.

Tom and Win had no children of their own and although life had been tough for them, they both shared a most infectious sense of fun and good nature. During their visits to various meetings, Win had been told she had the making of an excellent trance and materialisation medium. She often jokingly scoffed at the idea with the comment, "Can you imagine me?"

But, despite her disbelief, it *did* happen and Win developed trance mediumship which—despite the lack of education as a child—sometimes included masculine full voice conversation in fluent German and Latin! Often, Win used to laughingly chide Tom about sending her off to sleep so he could have a chat with another woman "on the other side."

As my wife and I lived at the other end of the country, we weren't able to share much of this activity. Then, on one occasion when paying them a visit, Tom mentioned that they had been promised that "when conditions were right" they would be given a demonstration of teleportation. I was naturally intrigued and went into some length discussing the possible 'mechanics' of this with many questions; what materials, inanimate or animate bodies, etc., etc. However, this was a rather one-sided conversation as Tom wasn't at all technically inclined.

Then several years later, when Win, Tom and myself had forgotten about it, the promise was kept.

Alone one winter night, seated on either side of a brightly burning fire, Tom and Win were having a nightcap before retiring. Suddenly, their attention was drawn to the ceiling by a 'tinkling sound.' Shadowy and vague at first, a small object materialised and dropped to the floor between them. Tom said that when he picked it up he found it to be a slightly warm little trinket. It turned out to be nothing more than a small brass bell and a little wooden cross, secured by a gaily coloured ribbon!

This promised demonstration had been for them alone and they treated it as such. In fact, we didn't learn of it until visiting them in 1962. It was a great thrill for me particularly, as it occurred privately to Win and Tom who, being like a family to us, we could trust implicitly; neither could they have possibly been mistaken.

I often pondered on the composite nature of that little gift. Why the variety of materials? Why not just a coin or something like that? Its material worth was nothing, yet in another way, priceless. Then years later—after our dear friends had both passed on—I was describing this little object to an acquaintance when suddenly I remembered Tom's quizzical look those years ago when I had, perhaps boringly, speculated with him on whether it was easier to teleport single substances such as stone or copper, and would composite structures make special problems? Then, in mid-sentence, I paused and wondered . . . my wife and I had seen and handled that little trinket which was treasured by my friends, but after a few years it vanished and was never found again.

These have been very precious first-hand experiences for me and I count myself as being very fortunate in having had them, for after all they are the kind which can only be matched by exciting finds in the laboratory. Yet how sad, if not strange, their kind should be so ridiculed and abandoned. However, if this type of phenomenon *is* accepted as fact, with little difficulty it too can be explained in terms of the Comat theory, for the very same reasons which were offered to explain levitation, dowsing, telepathy and homing instincts in animals, etc.

When giving talks on this aspect of the subject, I have been asked the question, "At what speed might an object be teleported, or would it be limited to the speed of light?" More often than not, out of sheer exasperation, I am tempted to seek solace in Hoffman's answer to the little boy's chicken and egg situation in Chapter 3: "er, yes." But usually, time permitting, I am grateful to offer my timeworn—hopefully useful—simple analogy.

"Beam Me Up, Scotty"

Consider particles in a structure represented by modulations on a carrier wave. In Chapter 5 it was shown how what we call motion may in fact be a kind of modulation transference due to 'resonance.' This being so, it is a short step to understand that vibrational phenomena of this kind can be transferred from one grid reference in the Matrix to another. The 'trick' being to sympathetically reform all the modulations along the Comat waves uniformly—or analogously—as follows.

Fig. 3.7 (a) depicts a very long plastic tube containing a fluid—such as water—representing a 'C' ray as in Chapter 4. At point A there is an inflated portion on the tube, representing a modulation (particle) which we wish to transfer to position B which may be, say, a thousand miles away; here we have to neglect such encumbrances as friction and weight, and remember it *is* only an analogy.

At B there is formed a local weakness in the tube, so that if the portion at A—the 'particle'—is squeezed it ceases to be, i.e., it vanishes, while in unison with this an identical 'particle' appears at the weakened portion at B.

In this analogy we assume water is incompressible, therefore the time taken for the *materialisation* of the new 'particle' at B, is exactly that required to cause the *dematerialisation* of that at A, let us say this might be .1 second. Suppose the *actual* longitudinal displacement of the volume at A to A1 be—say one inch—then the velocity along the tube would be around .56 miles per hour, as in (b). Whereas to *bodily* move the *initial volume* at A along the tube to point B—i.e., 1000 miles—in the same time of .1 second would require moving it at a velocity of no less than 36,000,000 miles per hour! [Fig. 3.7 (c)]

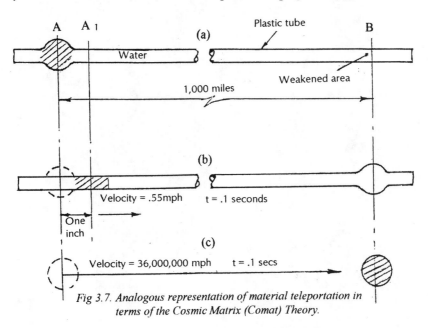

Fig 3.7. *Analogous representation of material teleportation in terms of the Cosmic Matrix (Comat) Theory.*

This is, analogously speaking, the case for Demat and Mat, i.e., in real terms, perhaps *instantaneous* transference of specific Comat modulations in time and space; again, it will probably take comparatively small amounts of energy to do it. Moreover, if it can be accepted that such phenomena *do* spontaneously occur in nature, then as with gravitation it should be repeatable—perhaps electronically.

Although diehard sceptics may scoff at the proposition for teleportation, the fact of the matter is that such phenomena *really do occur*. My primary task at this juncture is to bring attention to the fact that this is easier to accept when it is viewed within the context of one *overlaying cosmic principle*.

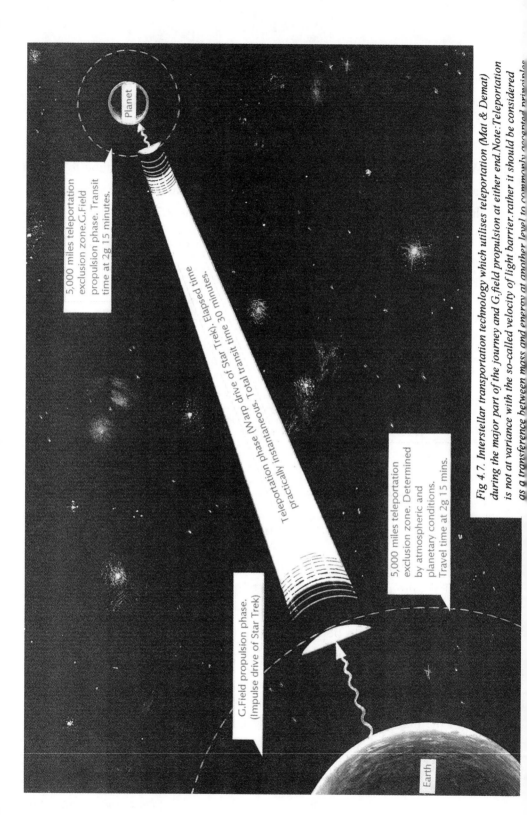

5,000 miles teleportation exclusion zone.G.Field propulsion phase. Transit time at 2g 15 minutes.

Planet

Teleportation phase (Warp drive of Star Trek). Elapsed time practically instantaneous. Total transit time 30 minutes.

G.Field propulsion phase. (Impulse drive of Star Trek)

5,000 miles teleportation exclusion zone. Determined by atmospheric and planetary conditions. Travel time at 2g 15 mins.

Earth

Fig 4.7. Interstellar transportation technology which utilises teleportation (Mat & Demat) during the major part of the journey and G.field propulsion at either end.Note:Teleportation is not at variance with the so-called velocity of light barrier.rather it should be considered as a transference between mass and energy at another level to commonly accepted principles.

Unfortunately, one-sided bias towards such phenomena is not entirely the myopia of the layman, for some otherwise brilliant cosmologists and writers now and again reveal that they are not without a degree of tunnel vision themselves. This author knows that teleportation occurs, as do many time-experienced researchers and, as with any other subject, the information is available to anyone who has a wish to find it. But it will not be learned by paying lip service to untrained, self-opinionated 'experts.' I repeat, it is not without significance that there has *never* been one unbiased scientist who, having devoted *years closely* studying paranormal effects, has not ultimately been utterly convinced of their actuality. The same can be said with regard to UFOs. Verification should be left to the testimony of these people. It is totally unacceptable to expect to get a true evaluation from results obtained by a few sceptical scientists from relatively brief, unsuitable laboratory sessions. For what it is worth, my advice is, please go and seek for yourselves before making hasty public declarations, which will assuredly one day be revealed as biased balderdash! I have little doubt that many of the predictions made in this book will eventually be substantiated.

For instance, I have very good reason to believe the process for interstellar travel will involve the following procedure in which both G. field propulsion *and* teleportation are utilised. Fig. 4.7 shows a typical space journey from the Earth employing this technique in which G. field propulsion is adopted at liftoff through the atmosphere to a distance of around 5000 miles; note there is no orbital motion or kinetic release velocity required. This distance—or exclusion zone—is necessary, due to the fact that teleportation of a massive vessel within a dense atmospheric belt can create a violent aerodynamic displacement similar to an explosion, in addition to attendant adverse electrical effects. Concerning this, it may only be of passing interest to note that—as we shall see later—the term G. field propulsion could quite easily be described as a kind of 'impulse drive,' while there could be an excuse for labelling teleportation as *Warp speed*, both assignations being typical of the *Star Trek* series. Full marks for the writers of the series!

However true or false this may be, there can be little doubt that long before any of these exotic adventures, we shall be employing relatively crude nuts and bolts anti-gravity techniques. It has to be reemphasised that among those physicists who *are* currently experimenting with teleportation that they, too, are extremely handicapped without prior knowledge of an extended cosmic theory, including the presently little-understood paranormal.

Dimensional Change and Teleportation

Earlier in Chapter 3, we saw how we might gain an idea of the extreme tenuity of matter and how according to basic physics even the fundamental particles constituting matter can quite fairly be regarded as no more than states of energy. This state is more typified by noting that it is said that the electron has a rest mass of zero, in other words it has only inertially acquired mass so that, in theory, it should vanish when it is stopped. Therefore in this sense we are quite justified in expressing physical phenomena in terms of an overlying vibratory state.

Now the question arises, assuming that modulations forming 'particles' of matter can be 'dematerialised,' then 'rematerialised' as in teleportation and bearing in mind the foregoing spatial tenuity within matter, how are we to visualise a dimensional space reference for reconstitution? In other words, what determines the reconstituted size of a rematerialised physical structure?[4]

Indeed we might add, could it be that such a process might be specifically employed for certain conveniences? For instance, a physical body such as a spacecraft together with its occupants, could be reduced to a fraction of its original size. Granted, at this moment in time, for us this can only be purely speculative, save to add that we shouldn't be surprised if such a phenomenon was observed in relation to a technologically advanced species who had perfected the art of teleportation, indeed *it is what we should expect!* For some readers I do not have to amplify the fact that in certain circumstances such a phenomenon is far from unknown.

It may also be significant to some that even a passing acquaintance with the plausibility of the foregoing deep space transport technique reduces current objections to deep space visitors to our planet just a little, when we reflect the objections are *based solely on limitations of our own space flight technique employing chemical rockets!*

A Kind of Time Travel[5]

At this juncture, I am reminded of the fact that it is difficult for many people to relate to seemingly enormous speeds, which after all is largely a relative factor. However it does help to remember that the visual effect of a car travelling at 65 mph in fairly close proximity to a stationary observer, is reduced to a snail's pace when viewed from above at several hundred feet in a stationary helicopter. Indeed, if it were possible to be situated in space within equal viewing distance of both the earth and the moon, a spectator would have to stay in that position for around three days watching a moon rocket travelling at an average speed of approximately *3-1/2 miles per second* across that void—that is if the ship could be visible through a telescope at that distance. Even if the vehicle was G. field-powered—as developed in Chapters 8 and 9—and travelling at a turnaround 1g acceleration, it would require an observer to watch patiently for approximately 3-1/2 hours before the craft reached the moon, during which time it would have acquired a maximum velocity at the mid-point of *38 miles per second*, or around 137,000 mph, and the observer would have detected no more movement than the hour hand of a clock!

It is interesting to note that in *The Evidence for UFOs* Hilary Evans says:

> The question of where the UFOs come from is one in which an anthropomorphic attitude has been allowed to prejudice the argument. Sceptics object that almost any interstellar journey would require many light years, often many lifetimes. Others reply that to think in terms of a literal real-time space journey is naive. The alleged visitors from UMMO, for example, claimed to make use of 'folds' in space which serve as shortcuts; and even if the UMMO communications turn out to be hoaxes, they indicate the sort of possibility *we have no right to rule out*.

The Parallel Universe Theory

If a process of teleportation *is* ever realised, then in terms of the Theory of Unity there can be little doubt that the operational means will be electrical. Indeed sci-fi writers have been using this theme for many years now. Similarly, in order to explain some apparently precognitive phenomena, they have exploited the theory of parallel universes. So for me or anyone else to have glimpses of future events in either sleeping or waking dreams suggests that time itself may be manipulable, with a parallel universe as an operative factor.

In saying this I am not implying that Avenel's position was faulty, rather that it was necessarily limited. Today we might reconcile that part of the theory with the multiple overlaying universe concept becoming popular in more recent times.[6] In fact, at one time Avenel implied that his single Source in the original Unity of Creation article had been chosen for simplicity and *there are possibly other coexistent sources of Creative rays.* In which case it is interesting to speculate that the basic space forming of each may not necessarily be the same, therefore two space matrices might occupy overlapping regions at the same time, so that a particle 'A' could in fact be created in an identical position to that of particle 'B'—together with all the aforementioned extension modulations—and neither system would interact with the other. Such a hypothesis could inherently embrace the proposition for a parallel universe, not least we are able to view a working model for a basis on which to build the phenomena of precognition, time travel and teleportation, etc.

Along with many others, at one time even Britain's leading cosmic physicist, Professor Stephen Hawking of Cambridge University, denounced the possibility of time travel, but recently he is quoted as having "sent shock waves through the Universe" by suggesting that indeed it just *might* be possible. With respect the author *knows* that it is not only possible, but as long ago as 1952—and since—I have been personally involved with irrefutable evidence of it. Nearly seventy years ago in his world-renowned book *An Experiment with Time,* J. W. Dunne publicised his "Serial Time Theory." At that time some people opined that masterly as it was the theory lacked a certain continuity. I suggest that due consideration of the global basis of the Theory of Unity might offer a link towards bridging that lack of continuity.

As will be seen from some of the examples already mentioned in this book and more graphically in the Appendix—and to say nothing about the historic and day-to-day precognitive phenomena which continue to occur—we are overwhelmingly forced to accept some future events seem to be prerecorded. The following, necessarily brief theoretical explanation might be helpful.

From Time to Time

In Fig. 5.7 the dotted lines represent only a few intersecting 'C' rays and the two heavy parallel lines represent two halves of one 'event' line lagging one behind the other. The dots P_1 and P_2 represent one person also split into two parts, and consequently they too are lagging one behind the other.

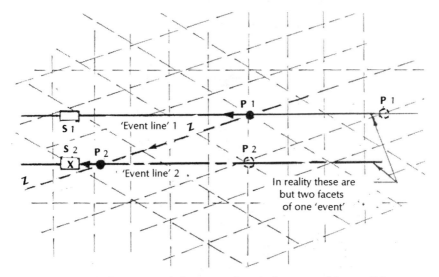

Fig 5.7. 'Precognative' phenomena described in terms of the parallel universe hypothesis.

The two 'halves' of this 'person' are travelling along a railway—depicted by the heavy 'event' lines—towards stations S_1 and S_2. Thus by equipping person P_1 with special 'viewing glasses' he could look along the intersecting line (creative ray) ZZ and see person P_2 (in reality himself) meeting Mr. X *before* his own 'event' occurred and in that sense he would be "blessed with foresight"!

This oversimplified example represents only one 'event' split into two halves in a 180 degree out-of-phase continuum, but like Dunne, Bond and Einstein we have to imagine that such multiple universes are infinite, in which every alternative variant could occur, i.e., in some instances Mr. X wouldn't be at the station to meet person P_2 and so on.

Invisibility

Throughout history there have been cases involving invisible assailants, while many devotees of some mystics claim they had witnessed the process of invisibility.

From purely optical considerations it is theoretically not difficult to fool the human eye, as is obviously the case with some professional illusionists. In fact many years ago when I was engaged as a consultant by Messrs. Lesney of Matchbox Toys, I was delighted to produce a toy in which a revolving miniature

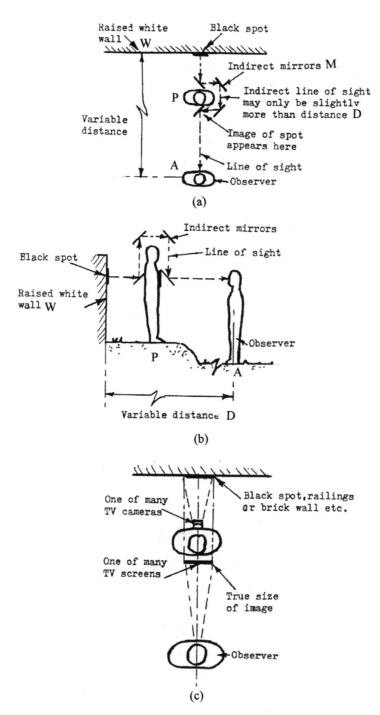

Fig 6.7. Elementary step analogy of invisibility created by
superimposition of an image in front of an object which
would normally be obscured by the object.

car was rendered invisible. But the difference in this case was the fact that I was able to produce the illusion in two stages, in which the outer body of the small car became transparent first leaving the internals—chassis, engine, suspension, wheels, etc.—clearly discernible before finally these also completely vanished. Of course this was rendered more difficult by the fact that the viewer was able to look all around the revolving exhibit.

I often remembered having seen this effect somewhere many years ago, but couldn't remember the details save that the exhibit was a radio set in a shop window in London. Then, many years later, while working for Lesneys one evening I happened to look out the uncurtained workshop window and saw my reflection apparently sitting at the wheel of my nearby parked car! The effect was very real but was only made possible due to the fact that the distance, height and posture I happened to be in, in relation to the window was exactly the distance, height and required orientation of the parked car. But of no less importance was the fact that the lighting also was absolutely right; subdued room lighting with a brightly lit anglepoise bench lamp locally illuminating me and the car outside. Within half an hour or so I was able to proudly display a partly opened full box of matches on the table and, much to the consternation of my young family, caused the matches to disappear! I was to learn much later the trick was as old as the hills, but after all those intervening years I had jolly good satisfaction in achieving it!

While further description is uncalled for here, in keeping with the main objective of this chapter it is worth noting that common optical considerations reveal the beginnings of a case for apparent invisibility, as shown in the sketch Fig. 6.7 (a). This is almost self-explanatory in that a person P is standing near a white wall W on which is painted a large black spot. At position A stands an observer whose line of sight towards the black spot is interrupted by P. However by the simple expedient of arranging the four mirrors 'M' the image of the spot is reflected around P's body—much after the fashion of a submarine's periscope—and therefore viewed by A in exactly the same apparent position it would have been seen had P not been present.

There is of course a slight inaccuracy involved, in that the apparent size of the spot will now seem slightly smaller than it would have been with direct uninterrupted viewing, due to the effective increase in distance the light rays have to travel. However, it will be noticed the discrepancy becomes less significant as the observer's distance increases. Fig. 6.7 (b) merely represents a side view of the same situation employing a further array of mirrors to establish the vertical position of the spot.

Today there are scientists who are working on a variant of this relatively crude analogy by the supplementation of hundreds of tiny double action TV projector/cameras which will cover the entire surface of an object in such a manner as to photograph any background imagery otherwise obscured by the object and relay this back—without dimensional distortion—to any observer, as in (c). In the timeworn adage, "bend the light around" him, her or it. Now this won't be perfect but such technology isn't very far away and given its plausibility, and bearing in mind the aforeclaimed wave structure of matter . . .

Electric People: A PK-Triggered Legacy?

The following case and others similar to it don't include levitation, but it will be apparent that the electromotive force which appears to be involved is supportive of the overall scenario so far. In particular it will be shown that such PK effects are sometimes extended to machinery, but it is significant that this only seems to occur when there are people involved. In other words, there are few—if any—cases where the phenomenon has been caught on film without the presence of human operators.

Not only do such phenomena occur, but it has been since recorded history. Indeed those individuals who were thus afflicted were often labelled the 'electric people.' One of the earliest scientifically investigated cases of an 'electric' person was that of Angelique Cottin of La Perriere, France, whose strange condition began when she was fourteen, on January 15th, 1846, and lasted for ten weeks [Fig. 7.7]. When she approached objects, they retreated from her. The slightest touch from her dress or hand was enough to send heavy furniture jumping up and down or spinning away from her. No one could hold an object she was also holding without it writhing from their grasp. A study group was appointed by the Academy of Sciences, and Francois Arago, the famous physicist, published a report in the *Journal des debats*, February 1846. He noted that her power appeared to be like electromagnetism—compasses went wild when near her—was

Fig 7.7. From an old print of one of the earliest scientifically investigated cases of an 'electric person' was that of Angelique Cottin of La Perrière, France.

stronger in the evening and seemed to emanate from her left side, particularly her left elbow and wrist. Poor Angelique would often be convulsed while the phenomenon was active, her heartbeat rising to 120 a minute and she was so frightened by it that she often fled from the scene.

More recently, in February 1996 a programme was shown on UK television presenting Telekinetic and Psychokinetic effects. This included close-up photography taken under test conditions before scientists, of a demonstration by Jean-Pierre Girard in which he successfully moved objects by a process of intense concentration.

A similar feat performed by Nina Kulagina of Petersburg was shown, in which she was able to move objects consisting of a variety of substances. Nina is

Fig 8.7. Despite three complete wiring installations of his home Frank Pattimore's electrical life is a misery. TV sets, lamp bulbs by the score.and every other electrical appliance are caused to explode by strange surges of electricity sometimes exceeding several thousand volts   (a)

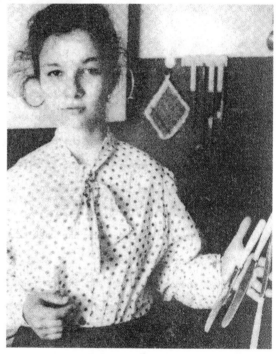

The strange power of Inga Galduchenko has attracted the interest of scientists worldwide.   (b)

also on record for producing levitation of small artifacts under close scrutiny of cameras, etc.

These cases are typical, but it is astonishing that so little, if any, research has been done even at the pathological level. Among those who *are* devoting their energies investigating paranormal phenomena, many are inclined to intuitively feel that there is something missing within the framework of their investigative abilities. With respect, I suggest that something might be an acceptance of the Cosmic Matrix hypothesis. Fig. 8.7 (a) and (b) illustrate some more recent typical cases.

A Psychokinetic or Bioplasmic Detector?

Past experience has indicated that some readers may wish to participate in the following simple intriguing experiment which anyone can repeat with a few minutes' careful preparation. It first came my way several decades ago and I understand it was investigated by several eminent physicists, including Sir

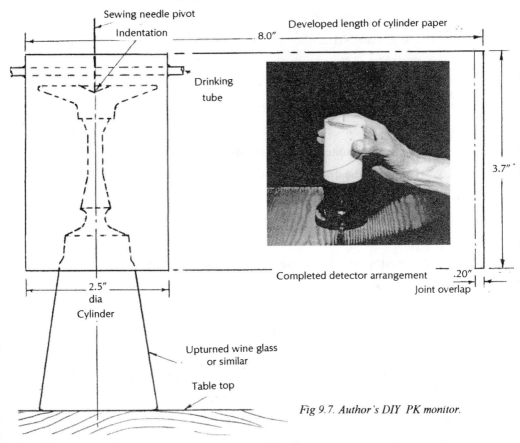

Fig 9.7. Author's DIY PK monitor.

William Crookes.

The apparatus consists of a small paper cylinder which is diametrically pierced at the top end to take a short length of drinking straw or tube. A small sewing needle is pushed through the centre of the tube so that its point protrudes about a quarter inch or so. When the cylinder is placed over an upturned wineglass, the point of the needle is lowered onto it, forming a practically frictionless bearing which can be moved in several axes to find the best balanced point [Fig. 9.7].

The operators should be comfortably seated in a draught-free room with the apparatus in front of them. The hand is then placed with the palm slightly cupped, close to, but not touching the cylinder, while the free hand can be placed over the nose to prevent air eddies due to breathing. Then all that is required is patience—and a little belief! Slowly, at first, the cylinder will begin to rotate until it gathers quite a speed. The rate of revolution will vary from person to person; for some the cylinder will not revolve at all. It should be noticed that it will always rotate *away* from the operator's outstretched fingers. This will be so for the left or right hand.

Having experimented with this device over the years, I can, of course, predict some of the explanatory observations, ranging from rising air currents from the warmth of the hand to uncontrolled breathing turbulence, 'electric wind,' etc. Some researchers have reproduced cupped latex gloves filled with body heat water, but the results are not the same. Moreover, the interested and patient experimenter will observe things which took us a long time to establish empirically. For instance, the speed of rotation not only varies from person to person, but it can vary for one individual as well. There appears to be a link between one feeling tired or energetic. Even more interesting and perhaps significant, the speed of rotation can vary with the geographic positioning of the operator, i.e., compass-bearing relationship.

There will be times when the device appears to become *polarised*, in that it doesn't 'know' which way to rotate. In this instance the operator must move away from the instrument for a while, after which the operation will return to normal, i.e., away from the fingertips. Another variable to try is with two people taking part, one with outstretched palm pointing at one side of the cylinder and the other person doing exactly the same thing on the other side. The rotation can thus be stopped by the two operators pointing—but not touching—their fingertips towards each other on the same side of the cylinder. Then, if it happens to be male and female taking part in this experiment, rotation may recommence, the direction of which seems to vary as if one of the participants is radiating *energy* and the other is receiving it. Normally with male and female the direction of the rotation is consistent, perhaps the reader may wish to find out. Suffice it to add that when there are two such participants, one old and the other young, or one in good health and the other poorly, the *direction* of rotation would appear to signify a giving and taking of energy, which could be bioplasmically engendered.

No less intriguing and perhaps of more significance is the fact, as one gets used to—or in tune with—the device, it begins to feel as though a certain amount of will comes into play, so that by intently imagining the cylinder's speed to increase or slow down, for some individuals there seems to be a response. Indeed I used to keep one of these gadgets under a domed glass cover on a shelf in my bedroom. On one occasion, while lying in bed recovering from a bout of 'flu, I was gazing at the little instrument after I had been pondering on some of the above-stated phenomena. I remember I was in a relaxed, semi-somnolent comfortable state with eyes unfixed and out of focus, when, with a start, I thought I detected movement. Somehow I *knew* I had to retain this position and dispassionately watch (as a bystander) as the cylinder gradually started to revolve, just as it had done on those many occasions when I had sat with elbows on the table trying to understand what was going on. But in this instance, I was separated from it by at least eleven feet!

It will be natural for some readers to suspect that as I had been unwell, perhaps I was a bit delirious; that was not the case however. As I record these facts, I know with certainty that after the publication of this book, there will be many people who, having constructed this gadget, will find that they are able to do exactly the same thing and far more successfully. The phenomenon just described is technically known to researchers as psi-induced telekinesis.

All sorts of control experiments can and have been tried with this simple little—what I like to call—PK turbine, including screening, temperature, electrostatics, etc., involving much tedium but no less fun, though of course at this time we cannot be certain as to the exact cause. However such seeming oddities suggest there are grounds for broader analysis when trying to evaluate some otherwise screwy physical results such as that obtained by Professors Fleischmann and Pons, who in 1989 stunned the world by announcing "they had harnessed the power of the sun in a test tube of water"—a discovery which would have led to cheap, pollution-free energy forever. But hundreds of leading scientists in America, Britain, Japan and other countries—who spent more than [British pound sterling]60 million in efforts to replicate the experiment—failed. Fleischmann and Pons were accused of "inventing data, breaching ethics, violating scientific protocol and fraud." Having quietly left the University of Utah, scientists were uncertain of their whereabouts. Eventually they were traced to Sophia Antipolis, 'a Science Park' twenty miles from Nice, where they are being funded by 'Technova,' a Tokyo-based think tank funded by Japanese companies including Toyota.

Dr. John Huizenga, co-chairman of a US panel of 20 experts (which failed to substantiate the Fleischmann and Pons claims) said their findings had been widely discredited by the scientific world. However, the former Southampton University Professors still say: "We have never participated in fraud nor invented data, and our experiments were carried out in a proper manner." A legitimate question has to be asked: could it be that the frustrating results, obtained by these scientists in their efforts to obtain fusion at room temperature, are bedevilled by extraneous runaway PK effects of one or both of them?

Corroborative PK Effects in the Laboratory?

It would be interesting to learn of the statistics among the various researchers who have noted a possible psychokinetic relationship with their work. It is revealed in the Appendix that John Keely was certainly acquainted with this phenomenon, as were Hendershot and others. Here again it would be remiss of me to omit a case with which I was involved, especially as it entailed supportive ramifications for the important developments introduced in this book.

Ron Howse and I first met when I joined the Aero Engine company D. Napier & Son in 1948. He is an inventor, a brilliant design engineer, lectures on astronomy and formed his own astronomical society and, despite the fact that we only get to meet once or twice a year, he has remained a close friend to this day. Naturally, therefore, we share interests and have endless discussions when we do meet.

In those earlier days, we pursued exhaustive excursions into dynamics, dreaming up ideas centred around Centrifugal and Gyroscopic space drives ad nauseum and whereas I have been inclined to transgress somewhat into the physicist's domain, Ron has primarily remained loyal to his original calling of mechanical engineering science.

In 1978, he was employed by an internationally known company in the southwest of England, who were manufacturing components for the much-publicised British Rail Tilt Train concept. One of Ron's projects was a revolutionary design for the coach bogey wheels main bearing housing (patented in his name). In order to fatigue-test this unit, he had to design a test rig to simulate the bearing running loads, only sped up to weeks, representing months of wear. This rig employed special transducers and servo units.

Over the years, Ron had nurtured a hunch that a simple spring biased mechanical oscillator might be encouraged to provide a reactive thrust force and, despite my oft-repeated Newtonian-based protests, I failed to convince him. No doubt having the gear ready to hand was an encouragement, for over a few lunch breaks he built a device and with the help of an electrical engineer, converted a solenoid (special electromagnet) to take a high speed make and break contact. This was mounted on a suitable frame and placed onto a console bench top for electrical checking, using DC current. Then they switched on the power, which was immediately followed by a loud crack. In the following silence they looked first at the bench top, then at each other in startled disbelief, for the contraption had become detached from its power supply, shot across the laboratory and smashed into a breeze block wall!

The unit was damaged, but not irreparable, so they rebuilt it and tried again, and again, but beyond the expected high speed vibration and loud hum, nothing untoward happened.

Ron eventually phoned me and in his typical, almost nonchalant way, described the event, quite convinced that he had been right all along and ended up by saying unbelievingly, "I really must try that again someday!"

Naturally, I took his word for it, but concluded that there may have been a perfectly *normal* reason for the action. So I asked a few obvious questions as to the exact circumstances and conditions prevailing on the bench top, etc.; there *had* to be an explanation for it. So for a few days I practically forgot about the matter until I received a phone call from another friend, unknown to Ron and who also lived on the mainland, with a message from someone (whom I neither knew nor met) urging me "to look again" at something extremely important with which I was currently involved. This also I shrugged off for a few days, then that nagging feeling now so familiar to me began again, and I knew I had to ask Ron more questions.

One night I phoned him and asked if he could please check out the following points for me: One, the weight of the unit. Two, the height of the bench top off the laboratory floor. Three, the distance from the end of the bench to the impact wall. Four, the height of the impact mark—also measuring from the floor—and Five, the input current and volt ratings—which happened to give some 720 watts.

The following evening Ron rang to give me this data and the first thing I noticed was that there was little difference in the bench height and the impact mark on the wall, this despite the fact that the end of the bench measured some eighteen feet from it. So we had the weight (and in order to be generous I allowed for an inch or two drop in height), thus we had the trajectory, acceleration and force required, which when computed read out at a staggering *fifty pounds* or so, for an electrical input of a mere 720 watts—or around 51 lbs. per h.p.!

Ron had assured me the effect could not have been caused by a normal reaction to anything on the bench, which had a polished top, so needless to say, I *was* puzzled.

The nature of the strange message—although obscure—was nonetheless an interesting, timely coincidence. But there was little more Ron could tell me and that is where the matter ended, that is until he next came to visit me nearly a year later.

Inevitably the subject came up again, during which I suggested that Ron should take me through the whole episode which had occurred that day in the laboratory, but now I had a few more questions to ask. What, for instance, was the geographical aspect of the layout? Ron didn't need much prompting, so with a clean sheet of paper we re-enacted the whole scenario, from the size of the room to the aspect relationship of the bench, equipment, power supply, who did what, etc., etc., right up to the moment when they had turned on the power. As Ron related the event—prompted by many interruptions from me—the full remarkable story emerged, revealing other facts, leaving me excited in stark contrast to Ron's rather passive bemusement. This is what actually happened that remarkable day in 1986.

Crunch Time

The two men had prepared the apparatus, positioned it on the bench, hooked up the power supply and got ready to switch on the power. At that point, I established that the electrician had been standing immediately in front of the unit, peering down at it with hands on the bench top, so that in order for Ron to reach up to the console to turn on the power supply, he had to bend over to the right-hand side of the electrician and steady himself *by placing his hand on the man's shoulder*. Normally that would have been of little consequence, but there was something else my arduous questioning revealed. In addition to the electrical contact points they had fitted into the rig—possibly for expediency—the electrician had hooked into a high frequency transducer line. This, together with the natural frequency of the solenoid's return spring . . . it would be anybody's guess as to exactly what went on. Little wonder they failed a repeat performance! I should add, there are good reasons to believe Ron H. has more than his normal quota of PK ability. I know that to this day he still believes he "did it his way," and I find it difficult to even begin to persuade him that, in all possibility, he triggered it—in effect—through the latent power within his mind.

During the past forty years it has been my good fortune to befriend other open-minded engineers and scientists from the scientific community, several of senior rank. One of them—Jack S.—while employed at the Atomic Research Establishment at Harwell, UK—also had one of these unusual tales to tell. In this instance, however, no special experiment was involved, rather an ordinary task of dynamically balancing an eight-inch diameter flywheel. Apparently, Jack and a colleague had installed the rotor (complete with shaft and bearings) into a rig and spun the assembly up by means of a special electromagnetic arrangement. No one to this day has been able to explain (though many have denied) how on earth it could be that this rotating mass could possibly completely detach from the stand, yet it did! Not only that, to the utter amazement of Jack and his colleague, it appeared to *casually* hover above the bench for an agonising instant before screaming off across the laboratory (unbelievably, the same as in Ron's case) before smashing completely through a partition wall. True, Jack said, the walls of the laboratory were only 'studding' but this 'missile' now went on through yet another similar wall before it left the building "by achieving a hat trick," but this time the outer wall was 4" thick brickwork!

Not surprisingly, the episode created quite a rumpus—and no doubt red faces—but nobody could explain it; neither could they explain why the three holes were at exactly the same height nor where the thing had gone, for it was never found again!

Perhaps understandably, the whole thing was hushed up. Yet the bizarre happening did have a positive side, insofar as Jack became more sympathetic towards 'fringe technology' than he might otherwise have been.

There was an intriguing sequel to this particular case when I visited Jack and his wife concerning a business matter. Just before Jack and I retired to his office, they had been telling me about the presence in their home of a deceased Rear Admiral who at one time used to live there. On several occasions they had seen him standing, looking out of the front bay window in the room which now served as Jack's office. He had seen the figure—as he said—"as clearly and defined as I see you now" while his wife said the figure was semi-transparent. At that time, they owned two large dogs who also seemed to sense the presence.

On returning to Jack's office that afternoon, he sat behind the desk facing the bay window and I sat at one side between the window and the desk. Suddenly, I noticed Jack was leaning back in his chair, fingertips to his mouth with—I thought—a pensive look on his face in reaction to something I was saying about the Admiral. I soon detected that he had obviously been distracted by his fax machine, which was operating. As the room was stacked with all sorts of hardware from videos, fax machine, photocopiers, etc., I had at first paid little attention to this noise above the rest. Then finally, Jack stood up staring fixedly at the fax machine, just a yard or so away; he looked at me incredulously and said, "That can't possibly be . . . the damned thing isn't plugged in!"

There are, of course, many other cases of normally electrically operated equipment functioning while disconnected. Several are known to me, such as my daughter-in-law together with a friend watching in amazement as a kettle of water was brought to the boil before it was plugged in, and an elevator which, under its own volition, used to operate before the petrified gaze of workmen who were demolishing the Southampton Palace Hotel in 1969. Both properties concerned had a reputation for poltergeist activity, which is frequently the case where electrical machinery and apparatus are involved. As was the case where the bewildered reporter's tape recorder mysteriously jammed—for instance—at the haunted house in Enfield we saw earlier on.

The foregoing examples of quite ordinary dynamic effects, associated with the paranormal kind, in which gravity and passage of matter *through* matter (Mat and Demat) were components, suggests a transference of energy into another—as yet—unknown force.

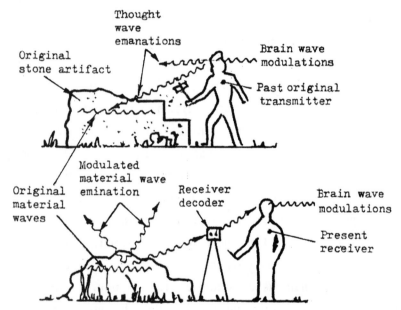

Original material forming waves are modulated by a past 'energetic' event which is detected by a receiver/decoder and relayed to the thought modulations of the percipient.

Fig 10.7. A mechanistic version of clairvoyance in which past imagery is replayed in the form of a 'ghost'

The Past Recorded?

This all too brief preamble around the halls of the presently labelled paranormal would be incomplete without at least a mention of the attempts of some more enlightened modern physicists at trying to explain some of the so-called ghosts and apparitions previously listed. Nowadays, it's not uncommon to hear physicists speculating on the possibility of certain emotive events somehow saturating surrounding physical structures—notably buildings and certain land areas—with 'energetic radiations' which permeate matter at the atomic level to produce a kind of physical record, similar to modern reproduced sound and visual techniques. As they see it, the chief difficulty will be in devising suitable electronic equipment to achieve this, which is unavailable at present. Here again, I suggest a full assessment of the wave nature of the Theory of Unity will prove to be of great assistance in this respect [Fig. 10.7]. That being so, the future of realistic appraisal of past events is exciting indeed. For instance, the mind boggles at the prospect of being able to realistically reproduce the actual building of the pyramids, while Jurassic Park would take on a new significance!

There are various convoluted processes by which some parts of the Comat can be liberated; such as interaction between energised matter and inert matter and/or by application of electrostatic or electromagnetic forces produced by coils, etc., which are all secondary effects, but in addition by the *direct* application of consciously stimulated primary forces within matter. Occultists call this psychokinetics, while some paraphysicists may prefer to designate the phenomenon as *Bioplasmic*. I suggest that most of us, under certain circumstances, have a propensity to operate as a kind of Biorhythmic Crystal.

It is supportive corroboration like this, together with my own experience and the known established paranormal effects—totally encompassed by the Theory of Unity—which has, over the years, lent wings to the strange exciting ideas I first had when, as a youngster from a poor family, a Mars bar advertisement kindled within me no less thoughts about the stars than a sweetmeat!

On a lighter note, past experience has also persuaded me that due to much of the foregoing, some not so well-intentioned members of the media will have a field day by accrediting this author as having stated that the world of the future will be powered by energy derived from poltergeists! But I am equally convinced that inclusion of such mechanistic anomalies here may promote a veritable avalanche of other similar, unrecorded instances, involving not only electronics and machinery, but *people*. Such spontaneous selectors of PK interference are anything but consistent. I cannot, therefore, be certain of my suspicions in connection with the extraordinarily temperamental results which have bedevilled the room temperature fusion experiments of the bewildered scientists Fleischmann and Pons, who stunned the world in 1989 when they announced they had harnessed the "power of the sun in a test-tube of water"!

Social Restraint Versus Technological Advance

While there are far more copious accounts of teleportation and other paranormal phenomena which help to focus attention on the general theory, those that have been quoted—including cases with which the author has been personally involved and can vouch for—are chiefly for the benefit of the unacquainted. In setting this down, I am acutely aware of the not so distant time when, if this hypothesis is correct, teleportation will no longer be the sole province of television magicians or mystics. What kind of society would we have to be in order to cope with a technology which implies that at the touch of a button we could virtually be somewhere else? The house of a friend perhaps, with prior arrangement would be nice. It would certainly cut down on the present-day fuel, pollution and traffic problem! But that capability would require that we were *all* socially orientated to a degree where we didn't indulge in curiosity about other people's private lives, so that unwarranted intrusion into our neighbours' houses would not occur![7]

How many people today could, with hand on heart, swear that they would be able to refrain from the tempting thought of being able to help themselves to just a little advance cash in order to help pay next month's bills, or a new car, better house, or perhaps a peep into a business competitor's file in order to obtain a nice fat contract? The truthful answer is, of course, very, very few.

Now that is accepted, so we will have to legislate the technological hardware. "Sure," I can hear someone saying, "just like CB radios were," and of course they would be correct. The technology might prove to be so simple it might be reproduced world-wide overnight!

In 1953, when I was first acquainted with the Theory of Unity, I soon visualised these possibilities, and at meetings I often proffered an opinion that in order for a society to continue, it would be necessary for its cultural excellence to develop in step with its technological development at the exponential rates predicted in Chapter 2. In other words, these two aspects of society *had* to develop hand in hand. At the time, I really thought that this *could* happen and we had a sporting chance; sadly, now, four and a half decades later, it looks as though my optimism was ill-founded and that *in this sense only*, some sceptics are correct. For this effectual peep into the future forcefully demonstrates our total incompatibility to accept or integrate with such advanced technologies, which let us have no doubt, are not only predictable, but much, much closer to actuality than the reader might imagine.

Nonetheless, as stated at the beginning of this book, it is quite clear, in order for our species to meaningfully survive at the *present rate of expansion*, we may have little alternative than to accept such dramatically changing technology, along with the attendant impasse.

Beyond any doubt, there would be enormous advantages to be had, particularly in the medical sector. That some kind of initial legislative control would be necessary is obvious and doubtless there would be an implied high risk aspect anticipated. In much the same way in which our forebears might have been horrified by the predicted prospect of visiting one's family *occasionally* in an automobile at 25 miles per hour, let alone at 75 mph *daily* in order to get to work! There is no reason to believe the teleportation situation would be any different in this respect.

In these last two chapters, we have been merely scratching the surface of some true facets of nature which, unfortunately, many are more comfortable to ignore. I submit that if we are to extend our present knowledge of the universe about us, we *must* be broadminded enough to address the paranormal. The author knows from experience that something along the lines of the Theory of Unity will help mankind to take that step. Granted, at the present time it may be difficult for some to build an identity between the paranormal and the well-established *normal* laws of physics. Very few physicists would dream of including discussions on such material as we have here. However, over the years I have witnessed signs of the predictable coming change in this direction, and it is hoped the foregoing may help to encourage an increased impetus towards that end.

In summary, I urge the reader to accept that there *is* a cosmic matrix and the *known electromagnetic spectrum* is merely a part of it. The remainder of this book represents an abbreviated account of work devoted to excursions into this unknown area, which was generally untrodden ground when I was still a young man. Today, some of this ground *is* being surveyed by a comparative few. It must be stressed, to what extent results of this work will be subtly influenced by the psi effect remains speculative at this moment. More strictly within the context of the main directive of the book, if, like the author, you have a scientific leaning and you are familiar with so-called paranormally induced levitation and teleportation as a *fact* of natural phenomena, then you *have* to accept that such effects should be duplicated by the application of intelligently directed manipulations of matter in the laboratory, as surely as man has emulated the flight of birds through the sky.

---

[1] There are cases where mediums at seances have been badly injured by similar intervention. In one instance, the medium was badly burned on the solar plexus when a bright *light was shone at her*. It is important that this and other effects are borne in mind as we proceed.

2 Including several successful business acquaintances of the author.

3 Materialised object.

4 We might presume initially this is correlated to basic wave length.

5 Many readers will recognise that the process of teleportation is inherently a means of time travel.

6 Einstein was the first to propose this idea. See also Appendix 2.

7 This may not necessarily have to be a transmit/receive process as we might currently envisage.

8

ELECTROGRAVITIC MOTIVATION
and Associated Phenomena
Including Crop Circles

This chapter may have significant importance, acting as a kind of narrow pathway between the 'borderland' sciences and the 'nuts and bolts' domain. At this time, the path is somewhat boggy and ill-defined, so as we are inclined to stumble around a little here and there, it pays to stop awhile and look back on the situation.

One thing that becomes clear is the fact that apparently very little in the way of energy and supportive technology is required to produce, or negate, paranormally induced levitation. Whereas, at the present time, even the advanced process now to be considered would *by comparison* require much of these. Clearly a closer evaluation of the former is indicated. As a first approximation, this can be achieved by another convenient analogy.

Consider the hot air balloon in Fig. 1.8 (a). We are acquainted with the fact that the balloon opposes gravity by the process of buoyancy which, in this case, is controlled by the addition or reduction of heat. But note, by the simple *physical* action of turning a stop cock which requires a relatively small force of a pound or so, the pilot can bring about a change in lift amounting to perhaps one hundred times that. In other words, given the energy—in this case fuel—the actual lift is dependent on the spatial environment.

In the case of the levitation of Father Dominic Carmo (b) the spatial 'environment' *was* the surrounding space, and the initiating 'fuel energy' resided in the living cells of his body. In principle, there is little difference. Thus it can be taken as a certainty, given the existence of a 'cosmic powerhouse,' that the process of levitation will be repeatable—albeit initially convoluted and perhaps comparatively clumsy—as I hope the remainder of this chapter will convey.

Another thing the last chapter makes clear is that gravity doesn't stand alone as an enigma! But even the apparently diverse nature of the rest of the

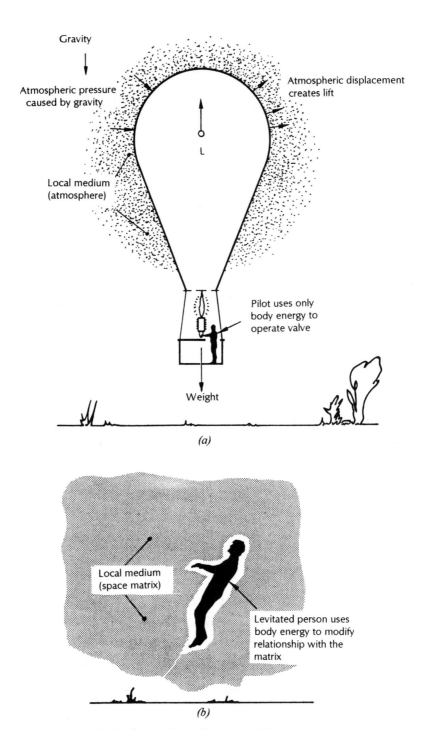

Gravity

Atmospheric pressure
caused by gravity

Atmospheric displacement
creates lift

L

Local medium
(atmosphere)

Pilot uses only
body energy to
operate valve

Weight

(a)

Local medium
(space matrix)

Levitated person uses
body energy to modify
relationship with the
matrix

(b)

Fig 1.8. Step analogy of 'paranormal' levitation

material becomes more acceptable when viewed in the context of an overlaying principle. Accepting this for the present, it is now required to correlate in-house some of the past issues, more specifically, the modulated wave analogies set out in Chapter 5. It was suggested that it is more convenient to view the situation in terms of rays 'passing through' matter, becoming 'diluted' and thereby creating a kind of deficit, rather than interference between waves. This still holds true. In order to develop the theory further, it is necessary to revert to wave phenomena language. This is quite permissible if it is remembered that we are still using analogical terms.

The following is the first of two accounts which, unfortunately, will not be found in classical physics textbooks. Students might have redirected their early thinking, had it been so. With the benefit of hindsight, we can see how much further the participants in these very important experiments might have gone had they prior knowledge of a cosmic theory such as the Theory of Unity.

Psychokinetic Versus Light Induced Levitation

In 1952, a well-known Viennese physicist, Felix Ehrenhaft, and his colleague, Ernst Reeger, proved that there is more than a little truth in the suspicion that tiny particles of dust tend to rotate when exposed to the rays of the sun. For they not only reproduced this phenomenon in the laboratory, but they succeeded in photographing it as well. To do this, Ehrenhaft placed graphite particles into a glass flask from which the air was completely evacuated [Fig. 2.8 (a)]. When the flask was exposed to focused beams of sunlight, as in (b), (c), (d), and (e), a large number of particles were seen to rise from the bottom of the flask instantly, and start to weave elliptical, circular and spiral shaped paths which were quite visible to the naked eye. The phenomenon ceased as soon as the concentrated light was weakened or cut off completely. Photographs taken at one fifth and one tenth of a second proved that not only were the particles orbiting, but perhaps more significantly, they were spinning on their axis.

It is of interest to note that Ehrenhaft wanted to relate the phenomenon to his own theory of a new type of physical force. He suggested that it is a purely *magnetic force* which permeates throughout *the known universe.*

It cannot be assumed that Ehrenhaft used the term *magnetic force* literally, he used it no doubt to save further and more complicated explanation, but there

Fig 2.8. Felix Ehrenhaft's dust particle test rig. (a)

(b)

THE DIRECTION IS SEEN WHEN THE CAMERA IS MOVED
he camera is moved. Instead of a circle we see a screw, showing that the
articles are rotating in definite directions. These screw courses seem to
emain in special planes in relation to the direction of the light-ray. The pearly
tructure shows that the dust particles are also spinning on their own axes.

can be little doubt that his theory will not be far removed from that offered in this book. The reader's attention is drawn to the fact that in this fascinating experiment, the particles are seen to have been levitated by the action of the sunlight, and although at first hearing this may seem to be at variance with the *negative* effect that light seems to produce in *paranormally* induced levitation, nevertheless there is a kind of parity in that there are three chief components involved: light, levitation and rotary motion; one result obtained with *physical* apparatus in the laboratory and another obtained by certain presently unknown PK processes in the room of a house in Enfield, as reported in Chapter 7. Readers will find that comparison between this and the following second account will prove significantly interesting.

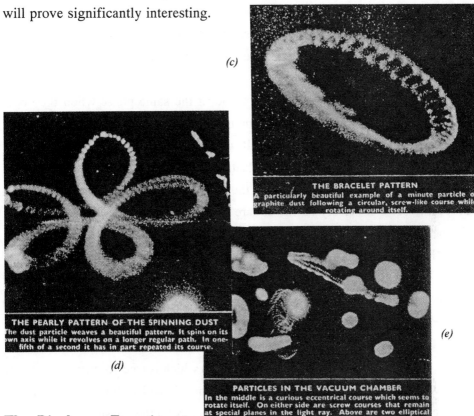

(c)

THE BRACELET PATTERN
A particularly beautiful example of a minute particle of graphite dust following a circular, screw-like course while rotating around itself.

THE PEARLY PATTERN OF THE SPINNING DUST
The dust particle weaves a beautiful pattern. It spins on its own axis while it revolves on a longer regular path. In one-fifth of a second it has in part repeated its course.

(d)

(e)

PARTICLES IN THE VACUUM CHAMBER
In the middle is a curious eccentrical course which seems to rotate itself. On either side are screw courses that remain at special planes in the light ray. Above are two elliptical courses uncompleted in the one-fifth second exposure.

The Bjerkness Experiment

At the 1881 Electrical Exhibition held in Paris, Professor B. J. Bjerkness of Christiana demonstrated some fascinating experiments, showing that it is possible to imitate most of the well-known magnetic effects by means of bodies pulsating in water. Interestingly, at that time, it was believed that the experiments offered a possible clue to the nature of the "mechanism which transmits electric and magnetic forces through space."

The apparatus used consisted of a glass trough filled with water, in which were immersed the small pulsating bodies. These consisted of little drums with elastic diaphragms at the ends, which were made to pulsate by drawing air in and out by means of pumps. The pumps had no valves and were constructed like a child's ordinary squirt, so that the air was drawn in and out at each motion of the piston. There were two pumps which actuated the pulsating bodies, whose mutual actions it was desired to study. These were driven rapidly by a hand wheel, and by altering the position of one of the crank pins, they could be made to work in the same or opposite phases. In the experiments, one pulsating body was held by hand and the other pivoted on a stand, so that it was free to move like a compass needle under the 'attraction' or 'repulsion' of the first one.

Phases

Two pulsating bodies were said to be in the same phase when they both expanded or both contracted at the same instant and in opposite phase when one expanded, while the other contracted. It was found that there is a close analogy between the mutual actions of the pulsating bodies and of magnetic poles and electrified bodies, but that in all cases the analogy was inverse. The force in all four cases varies inversely as the square of the distance between the attracting and repelling bodies.

Attraction of Light Bodies and Soft Iron

If a suspended body is disconnected from the pump, then a pulsating body will repel it in the same way as an electrified body attracts light objects, or a magnet attracts soft iron.

Attraction or Repulsion of Compass Needle

A body oscillating along a horizontal axis which is free to turn, will follow or fly from a pulsating body just as a compass needle will follow or fly from the pole of a magnet. The direction of the force depends on the phase.

Two Magnets of Unequal Strength

If two magnetic poles of the same polarity, but of which one is much stronger than the other, are placed a little distance apart, they will repel. But if brought near together, they attract, as the large one induces in the small one a polarity opposite to its own and stronger than its natural polarity. Similarly, two pulsating bodies moving in the phase which produces attraction, and one which is much larger than the other, will attract when they are a little distance apart, but repel when they are near.

Diamagnetism

Faraday suggested that many of the phenomena of diamagnetism may be accounted for by supposing all bodies to be paramagnetic, but of different strengths, and that the apparent repulsion observed with bismuth and other bodies is only due to the stronger attraction *exercised* on the *medium* in which they are *immersed*. It is probable however, that this explanation is not sufficient to account for all the phenomena observed. The analogous case in the Bjerkness experiments is that the actions on a body are opposite, according to whether it is lighter or heavier than the medium in which it is immersed. Bodies heavier than water are attracted by a pulsating body, bodies lighter than water are repelled. In each case we consider the body heavier than water as a type of diamagnetic body, remembering that all the phenomena are inverse; that the one lighter than water to be similarly the (inverse) type of paramagnetic body. Thus, the body heavier than water is acted on like soft iron, the one lighter like bismuth.

If two magnetic poles of the same name be placed a little way apart, a piece of iron will be repelled from between them; if they are of opposite names, it will be drawn in. Similarly, if two drums are placed a short distance apart, a body lighter than water will be attracted to the centre if the drums vibrate in the same phases, and will be repelled if they vibrate in opposite phases.

Lines of Force

Professor Bjerkness succeeded in tracing out the lines of force in the water, due to the various pulsating bodies experimented on, in a form which enabled him to compare them with the corresponding lines of force displayed by magnets and currents as traced by iron filings. The apparatus consisted of a heavy metal ball, supported on a stand by means of a light steel spring. When this was placed at various points, it oscillated along the direction of the wave in the water. A fine rod attached to the top of the device projected from the water, carrying a camel hair brush, which recorded the direction of vibration on the underside of a piece of smoked glass. A series of curves were obtained, which indicated that the lines of force between the pulsating bodies in water are exactly similar in form to those between converging electric, magnetic and electromagnetic forces.

A Mr. Strob has repeated Professor Bjerkness's experiments and reproduced nearly all the phenomena by means of sound waves in air. He caused air, vibrating by sound, to transmit the forces in the same way as Professor

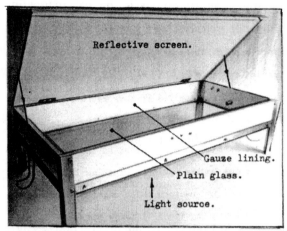

One of the author's ripple tanks (a)

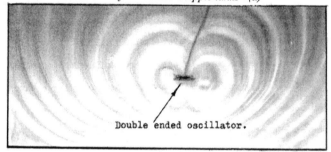

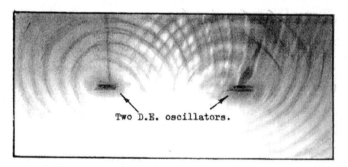

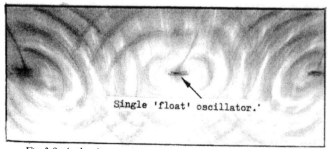

*Fig 3.8. Author's experiments showing wave patterns created
by phased oscillations in water. (b)*

Bjerkness transmitted them through water.

The summary went on to state, "In these researches we see opened a possibility of explaining some of the mysterious mechanism of electric and magnetic attractions, without the necessity of supposing any force to be at work other than those with which common experience makes us familiar, for we see them all reproduced by vibrations of material fluids, which differ from our supposed ether only in the superior quality of elasticity and the smaller density of the latter, that is, they differ from it only in degree not in kind."

The Extended Bjerkness Experiments

Today of course we use electrical equipment and a ripple tank to observe and photograph wave phenomena, so Professor Bjerkness was working somewhat at a disadvantage with the smoked screen and brush apparatus he employed to record the beautiful inter-reacting wave patterns obtained.

Fig. 3.8 (a) is a diagram of the author's modified ripple tank, while (b) shows some of the wave effects. It is well known to students of physics that there are two fundamental types of wave formations, known as transverse waves and longitudinal waves. All kinds of radiation: heat, radio, light, etc., are transverse waves which can function in a vacuum, while longitudinal waves are of the kind which transmit sound through a variety of substances—including fluids and gasses. This should be borne in mind as we proceed.

Among my original investigations of the Bjerkness experiment was a version in which a standing wave might be set up in water. In this experiment, two oscillators fitted with single diaphragms were secured diametrically opposite from one another with diaphragms facing. Provision was made for a third, normal, double-ended oscillator to be freely suspended between them, again with the diaphragms facing.

Nevertheless, It Does Move

The two fixed oscillators were set in the same phase to represent matrix wave intermodulations in one plane only, the third suspended 'float' oscillator represented one atom of billions in an inanimate structure. I well remember the occasion when the latter, vibrating at carefully controlled matching frequency, haltingly tended to move from one end of the tank toward the other, the direction of travel being *entirely dependent on the phasing*. Thus for the first time I witnessed a functioning model of what I considered to be a vibratory space-continuum in action. In this simple experiment, a physical device—suspended in a medium—was caused to be *displaced by the medium,* not in the fashion of a boat being carried along by a moving stream, or by means of ejecting some of the medium via an actuator to obtain thrust—as in classical mechanics—but motion in a governed direction due solely to synchronously engaging with an oscillating

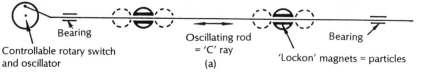

General arrangement and first phase in which both magnets are simultaneously and permanently locked on the oscillating rod, representing the 'normal' state of rays and particles

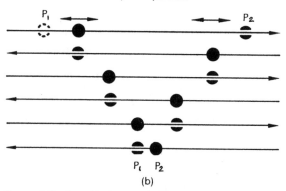

Sequentially phased lock on and off proceedure reproduces gravitational, electrostatic and magnetic 'attraction' phenomena while the opposite phasing produces 'repulsion'

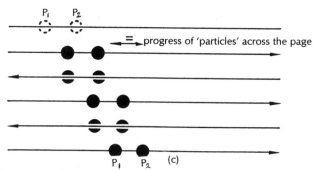

Independent and unidirectional 'thrust' of particles as produced in a gravitational field discussed in Chapters 8 & 9.

Fig 4.8. The author's mechanical sequentially phased lock on/off analogy reproducing the known physical interactions at a distance i.e. for the very first time attraction and repulsion phenomena are indicated as being one and the same thing.

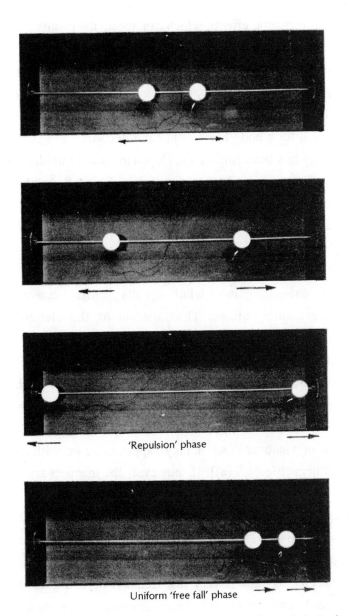

'Repulsion' phase

Uniform 'free fall' phase

Some action photographs of the oscillating rod and ball analogy, which simple action reproduces the main facets of natural physical phenomena , viz. 'attraction', 'repulsion', neutral and uniform motion in a gravitational field. Only two of these phases are shown here.   (d) medium. Naturally, sophisticated equipment employing modern piezo electric transducers are more accommodating, but care has to be taken to avoid any interacting electromagnetic effects, which can negate the results.

As stated, the analogy is limited to represent only one plane in the space matrix, nevertheless it does offer a visual measure of the fact that the energy in the oscillating medium might be *comparatively* large, and we have to imagine it being there all the time—free—it requiring only a relatively modest amount of energy to engage it for a body to move in any direction.

The analogy has been further developed to include unidirectional 'thrust,' in addition to 'attraction' and 'repulsion,' as in Fig. 4.8. In this, a steel rod, mounted on end sleeve bearings, replaces the water medium. The rod is coupled to a mechanical oscillator, which produces rapid and very energetic longitudinal oscillations of about 1.5 inches stroke.

Sleeved on to the rod are two, easy running solenoid-type electromagnets—encased in white balls for convenient visual observation—representing atoms. The function of the electromagnets is to powerfully grip the rod when they are energised, so that they are carried along with it. The electrical arrangement includes a rotary switch mechanism, which provides independent control of the magnets, as well as their phasing. In operation, the rod can be slowly or rapidly oscillated from end to end (representing a modulated 'C' ray) but due to their inertia, the two unenergised magnets remain in whatever position they happen to be in, with the rod slipping easily through them [Fig. 4.8 (a)]. If, however, the magnets are *simultaneously* energised, they lock on and oscillate longitudinally with the rod. But when they are *intermittently* and *synchronously* engaged—one on and one off sequence—they immediately move towards or away from each other—or unidirectionally together—as in Fig. 4.8 (b), (c) and (d), in short steps, which can only be detected as such at very low frequency. At higher frequency, it is impossible to detect the motion as being anything other than a smooth and continuous one. As stated, the magnet's directional mode depends *entirely on the controlled phasing* and it should be understood that this action depends solely on *mechanical* locking and is in no way influenced by the usual associated magnetic attraction or repulsion effects. In fact, but for the imposed mechanical difficulties, the electromagnets might just as effectively have been mechanical ratchets. By selection and balance of both the energy input to the magnets and the frequency of the oscillation, it is possible to achieve representative varying degrees of 'mass screening' by allowing a lesser or greater degree of magnetic *lock* and/or *slip*.

A Parlour Game Experiment

Occasionally, when my demonstration equipment was unavailable, I have sometimes in desperation (and no doubt to the amusement of an audience) resorted to a makeshift "on the spot" analogy of the foregoing process; readers might like to try it themselves.

For this parlour game experiment all you need is a pencil, a small weight (a book will do), several metres of string and three willing assistants. The string is passed over the pencil, held horizontally by assistant No. 1, the weight being secured to its lower end; this in effect forms a pulley and a return spring. The other end of the string is pulled across the room by the second assistant, who can raise or lower the suspended weight at varying speeds—to begin with, the slower the better! Thus a means is provided for the slightly tensioned string to be given a forward and backward motion of, say, 24 cms. Assistant 1 serves as the support bearing, the string represents the oscillating rod, while assistant 2 provides the motive power for the magnitude and stroke of the oscillations.

Having set up the 'rig,' this is where the fun begins! Assistant 3 stands facing the string—about halfway along its length—and carefully cups one hand around it, thereby completing the set-up, in which this assistant's hand represents one of the white balls in the previous analogy.

The parlour game experiment starts with phase one, in which the string is oscillated backwards and forwards at a *governed* speed, slipping through assistant 3's cupped hand—for, say, five strokes—but on the sixth stroke the string is grasped tightly allowing the hand to be pulled along with it. This represents the mechanical magnetic ratchet. The hand having been moved along by 24 cms then releases the string and allows it to slip back and forth for another five strokes before it is once again grasped and moved along a further 24 cms and so on, requiring assistant 3 to step sideways in *one direction only*. By changing the phase, left or right displacement can be achieved. For a constant oscillatory frequency, assistant 3's lateral displacement speed can be increased or decreased merely by grasping the string *in phase* with a *more* or *less* number of free slips. In the screening analogy, this represents varying degrees of substance 'transparency' or 'opacity.'

Having become proficient at this one hand trick—usually after a few minutes—the next and most important step can be taken, whereby assistant 3 stands with *both* hands apart and gripping the string. Then, in unison repeating the foregoing procedure, this time with both hands in synchronous motion together left or right.

By changing the phase of grip to a one on and one off sequence, the hands can be made to move towards each other in steps, faster or slower according to the frequency of grip, i.e., slip six or slip two, etc., as the case may be. This represents gravitic magnetic or electrostatic 'attraction'; moreover a little patience and experiment will show that by changing the phasing again, the operator's hands can be made to move apart, thereby representing magnetic and electrostatic 'repulsion.'

In a way, it may be fittingly appropriate that this—what I think may be a most important aspect of nature—should be so simply demonstrable by such humble means to anyone, anywhere. Not only does it duplicate the rod, ball and ripple tank displacement experiment, but for the first time correlates the main known physical laws in terms of Comat energy; i.e., according to their synchronicity and latching phasing the bodies will:

(A) Both move together as with magnetic, electrostatic and gravitational 'attraction' phenomena.

(B) Both move apart as with magnetic and electrostatic 'repulsion' phenomena.

(C) Both move in unison in a given direction—left or right—with unidirectional 'thrust' (as discussed later on in this chapter) and this due only to an oscillating medium; in this case a steed rod or length of string! It will be seen that the foregoing, functional analogies are in concordance with the modulating rays hypothesis set out in Chapters 4 and 5.

It is interesting to note that, if the rod were to be formed into a circle as in Fig. 5.8, it would be a nearer approximation to one modulated alternating 'C' ray, in accordance with the Theory of Unity. However, as with the ripple tank

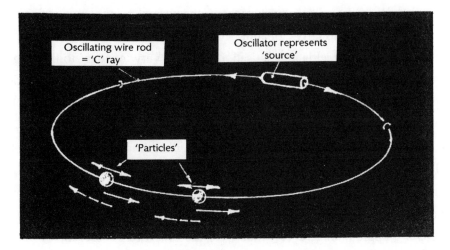

*Fig 5.8. Natural development of oscillating rod analogy to conform
with the Unity of Creation theory.*

*Note: If two particles are moving away-or towards-each other they are always on a collision
course when viewed from a source, signifying that the phenomena of attraction and repulsion
are one and the same.*

experiment, both analogies are restricted to representing only one plane. Even so, experiments like these have helped significantly towards the development of the following proposition for wave motivated and levitated vehicles. Suffice for the moment then, to imagine that the waves of space could be alternating and see what that might imply.

Electrostatic Opacity

There is, of course, a limit to the extent in which a theory can be represented in purely graphical terms, and it is anticipated that some readers will have quite a few questions to ask: Such as, "How do some other well-known aspects of natural phenomena relate to the general theory?" For example, how does it explain why iron is 'attracted' to a magnet, while a piece of wood is not?

As we have seen, the theory holds that all matter is subservient to the Comat insofar that it behaves gravitationally and it further anticipates that there are extension modulations of many frequencies, and the extraordinary electrostatic and magnetic properties of some materials are only two of these. Thus we might say wood is 'transparent' to a magnetic ray, while soft iron is not. This is in common with other physical phenomena, where, for example, sunlight passes through ordinary glass while accompanying ultraviolet rays do not.

Similarly—like all other substances—the normal arrangement of atoms in steel are only fractionally 'opaque' to rays of gravitational and magnetic frequency, yet when steel is magnetised, it becomes very 'opaque' to the magnetic part of the Comat spectrum and the same can be said of all electrostatically charged substances. As a first approximation, Fig. 6.8 helps to develop the principle a step further.

In this, wave energy 'dilution' and particle 'drag' terms—such as those employed in the earlier air jet and screen analogy in Chapter 5—are again retained. Fig. 6.8 (a) depicts two bodies having identical 'opacity' or mass, which we have for convenience designated as being 10 per cent. To the left of the body X there are approaching rays, which over the area of X represents a potential

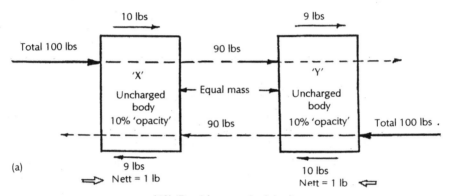

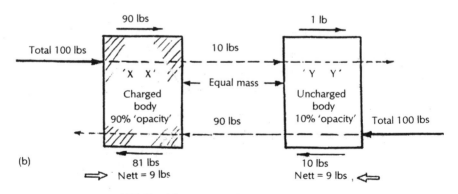

Fig 6.8. Two-bodied sequential step analogy toward an electrodynamic space drive.

force of, say, 100 lbs. These rays 'pass through' X giving up 10 per cent of their energy, which is now experienced by X in the form of a 'drag' or force of 10 lbs. towards the second body Y. But now these rays represent only 90 lbs. of force as they contact body Y which they now 'pass through,' giving up another 10 per cent of their reduced energy.

Thus body X has a gross force of 10 lbs. 'pushing' it to the right of the diagram and body Y has a gross force of 9 lbs. also 'pushing' it to the right of the diagram. Now, this represents rays in only one direction, but suppose they

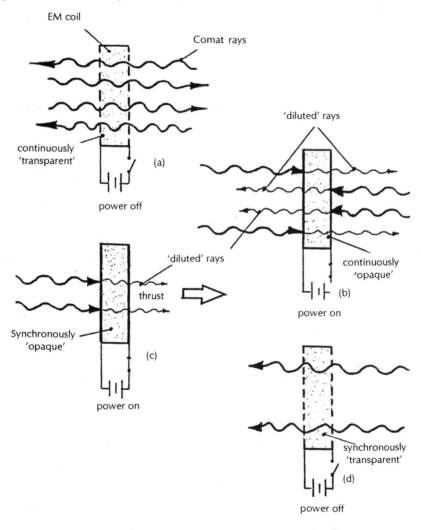

Fig 7.8. Single bodied step analogy schematic of an electrodynamically variable opacity drive unit.

could be analogously alternating? Then this same sequence must be inversely reproduced, i.e., from the right-hand side of the diagram to the left. A quick summary of this situation reveals that *both* bodies are therefore 'pushed' towards one another with a net force of 1 lb. each. This may be interpreted as the normal g force.

Fig. 6.8 (b) is a repetition of this two-body analogy. In this case however, although they both have the same mass as before, body XX is electrostatically charged and therefore has an increased electrostatic 'opacity'—of, say, 90%—while body YY still has the original 10% 'opacity.' It will be seen that despite XX having an electrostatic 'opacity' of 80% *more* than YY, the net 'push' is once again *equal* on both bodies at 9 lbs. 'push' each, in accordance with known laws (or approximately so).

In other words, this offers a correlation to increasing the closeness of the air jet screen weave portrayed in Fig. 13.5. Therefore, in terms of the analogy, it is important to realise that to electrostatically charge a body is to effectively increase the 'opacity,' or unbalance the extension modulations; in essence they are one and the same thing.

An Electrostatic Space Drive Analogy

We are now in a position to broadly develop the analogy to its logical conclusion with the aid of Fig. 7.8 (a), which again considers a body in an uncharged state and therefore 'transparent' to alternating Comat waves of electrostatic frequency and wave length. If it is now continuously charged, the body becomes 'opaque' to such waves and they, in turn, give up some of their energy to the mass as in (b), and in accordance with electrostatic phenomena, some of this dissipation is expressed as radiation—in much the same manner in which I believe planetary bodies may be internally heated, as suggested in Chapter 5.

In this *single* body situation, normally it wouldn't move, but as we are proposing, the Comat modulations are alternating—as indicated by the rod and ball analogy—motion due to the so-called 'attraction' which may in reality be a series of extremely rapid steps. If this and the 'opacity' factor are now correlated, we are one step nearer to completing a most important aspect of a true space drive representation.

In Fig. 7.8 (c) the body is electrostatically charged and is therefore 'opaque' and in phase with the right-hand directed waves. While in the left-hand directed phase (d), the body has been discharged and is therefore 'transparent' to the waves. In this manner, the body continues to move towards the right in a series of extremely short pulses which—depending on the frequency—can produce a virtually smooth continuous motion, in exactly the same manner as the rod and magnetic ratchet analogy (and, for that matter, as does a body 'fall' towards the earth)!

Thus, based on an all-embracing cosmological theory, we are faced with a simple proposition for a unidirectional space drive concept involving two chief attributes, in which a body is electrostatically pulse charged to provide an 'opacity-lock,' and these pulses *must* be precisely synchronised with the alternating frequency of the appropriate range in the Comat extension modulations. In other words, this basically simple arrangement moves, not towards another charged body which moves towards *it* and cancels out, but by 'latching' onto an extremely potent, monodirectional spatial displacement.

Due to the physical limitations (and some other unknown associated factors) exact frequency matching is of course highly improbable, therefore harmonic frequencies have to be assumed where 'hit rate' matching depends entirely on the elegance of the methodology employed.

Insofar as the theory is based on synchronising with the Cosmic Matrix, I suggest a more inclusive acronym would be 'Syncomat.' Thus we are talking about 'Comat' primary energy and 'Syncomat' technology.

It cannot be overemphasised that the power available is inherent in the *space matrix* and the 'deficit' (i.e., thrust) is only a measure of the 'opacity' and magnitude of the electrostatic pulses attained. It must also be emphasised, such a drive wouldn't be a 'gravitational' device such as that described later in this chapter, because it relates only to a specific fraction of the Comat spectrum. The occupants of a vehicle propelled by such an electrostatic thrust system would be subject to all the normal acceleration forces and this part of the overall hypothesis has now to be addressed, albeit in very broad terms.

It is well known to student engineers that when a hollow body is

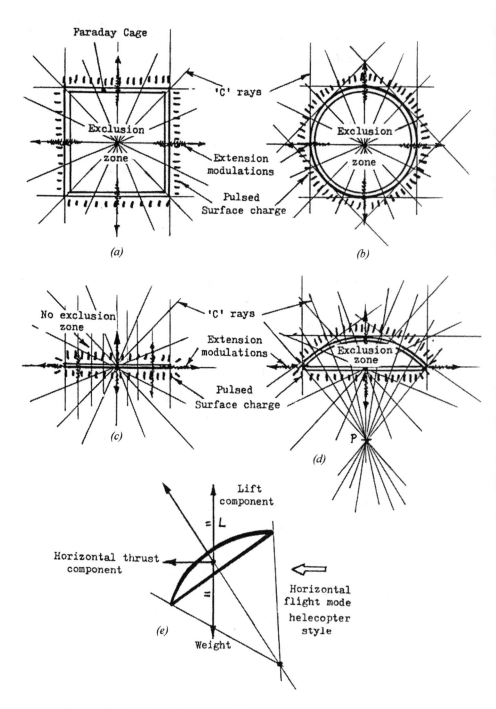

Fig 8.8. The logical development sequence for the shape of a synchronously pulsed motivated vehicle.

electrostatically charged, there remains an absence of charge within. This effect was first discovered by the eighteenth-century physical scientist Michael Faraday, and today one of its equally well-known derivatives is the 'Faraday cage,' which essentially is a cage capable of being highly charged while its contents (apparatus or people) remain completely isolated. A similar effect occurs if a ring of iron is subject to a strong magnetic field, the field tends to follow the perimeter of the ring leaving an isolated area in the centre. Both these phenomena are often used to advantage by electrical engineers.

Fig. 8.8 (a) is a charged cubically shaped hollow conductor, the radiating lines representing intercepting Comat rays in every direction. According to the foregoing analysis thus far, within and around such a body there exists a highly complex array of alternating extension modulations. Now although at this time we are not cognizant with the individual oscillatory interplay, nonetheless due to common observation we do have an approximation of the overall effects. Thus we can say that given the authenticity of a vibratory Comat, then it follows that should the above cube be synchronously electrodynamically pulse charged, any tendency for it to move in the direction of one side would be cancelled out by the adjacent side and so on. Although the cubically shaped example is easier to visualise, the same effect can be identified in the case of a hollow sphere as in (b).

Consideration of these two extreme examples suggests that we look at the other hypothetical extreme, in which a similarly synchronously charged body is represented by a thin flat plate; clearly any reaction which takes place is unlikely to be edgewise but predominantly in the direction of one or other of the major surfaces [Fig. 8.8 (c)].

It is now a short step from there to develop a working compromise between these two extremes, in which part of the hollow sphere, together with the flat plate, form components and we arrive at the typical flying saucer cross-section (d).

Further analysis of the direction of the ray forming modulations at local right angles to the two surfaces, produces a focus forming point P, which many years ago in *Piece for a Jigsaw*, the author tentatively identified as a 'point source,' which, at that time, I was unable to develop further. It will be realised that according to the premise so far, such a device literally becomes a directional motivating 'engine' in that—in a vertical thrust reaction mode for instance—a mere tilt from the horizontal, as in Fig. 8.8 (e), will produce a horizontal thrust component (exactly in the same manner I discussed at length in *Piece for a Jigsaw*). It must be restated that the occupants of this particular version of an electrodynamically motivated vehicle would still be subject to accelerating forces, despite the aforementioned cage effect.

Summary to the General Application

A brief summary of the foregoing analysis holds that:

1. There are sound reasons to believe that gravitational, electrostatic and magnetic attraction and repulsion phenomena are the same, having their origin in a Cosmic Matrix which is fundamentally of a wave character, according to the Theory of Unity.

2. The theory suggests that the basic wave modulations are alternating.

3. The tendency of thus formed particles to move toward or away from one another is due to an interference or imbalance according to the phasing. But this movement in reality is formed by an extremely rapid series of short vibrations in all planes.

4. Although the tendency to move toward or away is normally observed as a two-bodied phenomenon, under special circumstances motion can be initiated by the introduction of harmonically synchronised pulses of electrical energy.

5. In the instance of an isolated body, energetically pulse-stimulated so that motion takes place, the gravitational component is still active and remains undisturbed.

Completion of this step analogy brings us to the very threshold of real and potent gravitational manipulation. So, at this stage, it must be stressed the origin of this work is original and the findings are totally independent; any indication of similarity to the work of other researchers should reflect an encouraging uniformity, that is all.[1]

Levitation in the Laboratory

In Chapter 7, other cases of apparent zero gravity or levitation were cited—several of which would qualify as paranormally induced—together with the seemingly levitated graphite particles produced by Felix Ehrenhaft. In this instance, it is logical to assume this phenomenon is not due to buoyancy because the flask is highly vacuated, moreover the particles drop to the bottom of the flask the moment the sunlight is diminished and they are seen to rise the moment it is restored. Logically, it is difficult to attribute the phenomenon of rotating particles to an electrostatic effect and, clearly, insufficient information on detailed analysis is a disadvantage. However, we can make some very broad, useful comparisons.

We saw that, in this case, the levitating effect seems to be stimulated by the application of light, whereas paranormally induced levitation can be excluded by it. So despite this apparent inconsistency, clearly there is a relationship. We are justified in asking, is this merely a coincidence? While levitated, the graphite particles are shown to be rotating, as were clothing and other articles in the Enfield case, while a child was levitated. Is this yet another coincidence? These are only random, significantly interesting incidents involving light, weight and rotation. There are, of course, many more—some known to me. Suffice to add that among seemingly obscure phenomena, there can be found veritable gems of information for the unbiased enquirer.

Did Verschoyle Obtain Gravitational Transparency?

One of the earliest memories I have concerning the idea of gravity manipulation, is seeing a brief news item at a local cinema in London in the 1930s, in which a British inventor by the name of W. D. Verschoyle discussed some fascinating effects he had obtained. Some years after that, I vaguely remember reading an article by Verschoyle titled "Electrogravitic Lift." Then, later still, I met someone who was able to describe a demonstration given by the inventor. In this, a cigar-shaped model was suspended by a wire which passed over two pulleys mounted on the ceiling of the inventor's workshop, the other end of the wire being connected to a small counterweight and electrical apparatus. When Verschoyle switched on the power, the model rose from the bench up to the ceiling and descended when the power was turned off. I understood that the model contained a coil, through which the power was passed [Fig. 9.8].

At that time, some claimed that ordinary electrostatic effects caused the action, but I didn't feel comfortable about this due to the distances involved. Still, the rumours filtered through, and it was claimed that Verschoyle was employing pulsed charges. This seemed interesting, but I was finally very intrigued to learn that for some intuitive reason, best known to himself, Verschoyle had employed a fairly neat, variable input device.

As shown in the Appendix, Verschoyle vanished during a London bombing raid and has never been heard of again and we are left to reflect; if his suspended model—only partly weight compensated—rose to the ceiling and descended when the power was withdrawn, the phenomenon could only have occurred due to one

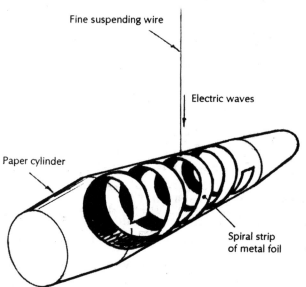

Fig 9.8. Verschoyle's 'Electro-Gravitic' device. It is significant to note that once most physicists begin to accept the idea of anti-gravity they immediately begin to think in terms of wave guides, nuclear power packs and anti-matter etc. Yet Verschoyle apparently obtained gravitational transparency as did saints like Father Dominic Carmo with the aid of none of these.

of three reasons. One, local electrostatic effects, i.e., repulsion, which distance would almost certainly rule out; two, reaction due to ejected ionised particles—which due to the horizontally suspended cigar-shaped model is difficult to reconcile aerodynamically—or three, and by far the most critical, that humble little paper and aluminum model became slightly 'transparent' to *gravitational* waves! This author finds it extraordinary that, at that time, such observations as these were available to anyone with an enquiring mind. Granted, Verschoyle's results—like his peers' before him (including John W. Keely)—were almost certainly purely empirically obtained. Clearly, this farsighted man was many years before his time, as I hope this book and time will show.

"Gravitational Transparency"

At this point, it should be understood that the term 'anti-gravity'—as in levitation—although perhaps not intended, does nevertheless imply an opposing force of the kind previously outlined, whereas 'zero-gravity' is a more passive representation of levitation which would be produced if a body was rendered 'transparent' to gravitational waves in *all* planes. As we have seen, inasmuch as gravity and inertial forces are one and the same, such a body could be moved in any direction by the application of an extremely modest independent force, as in the instance of Father Dominic Carmo Dechaux, which could only have been in the order of a mere *10 grammes* of thrust.

Assuming an average-sized person, and bearing in mind there is an absence of weight *and* inertia, the only resisting force would be due to atmospheric drag, so allowing a drag coefficient of about 1, indicates that Father Dechaux could have indeed been moved laterally at approximately one foot per second. In other words, the written testimony stands up to technical scrutiny. In such a circumstance of gravitational 'transparency,' it is debatable if the effect would extend into the air space around the body, but even if it did, the result would be much the same. However, the volume of displaced air its bulk represented would cause Archimedian buoyancy and it would tend to rise, in exactly the same manner as the lighter than air balloon discussed earlier.

Should the bulk be a vehicle of modest size, the lifting component might be a few hundred pounds, while if the bulk were that of an average-sized person, the lift effect would be only a few ounces and the rise would be slight, exactly in accordance with the instance of Father Dominic Carmo Dechaux. Unlike the foregoing electrostatic version, a body or vehicle suspended by such means, even if tilted, would be devoid of a lateral thrust component and would be isolated like a lighter than air balloon in still air.

When an Analogy Ceases to be an Analogy

Now, although the earlier synchronously pulsed electrostatic example is useful as a step analogy, it shouldn't be interpreted literally, because there are other processes involved which have to be incorporated.

It was stated in Chapter 4 that the frequency of the physical range of the modulated C rays is extremely high—perhaps too high to accurately measure—and the wavelengths correspondingly short, so it is in the area of microwaves which calls for investigation. But this demands too much from most available experimental equipment, so where does one begin? The answer to that question is simple; you do exactly what Verschoyle and other pioneer researchers did: you work empirically until you are lucky enough to locate "a signal among the noise." This may sound optimistic enough, yet it's not so optimistic if you are fortunate to have a good theory on which to work. That and a little foresight helps!

Verschoyle may not have had the benefit of a cosmological theory, yet he must have had plenty of foresight. It must also be pointed out that in the 1930s, many people were experimenting with home-built radios, the work of Nikola Tesla and wave transmission generally. In fact, it was claimed that when describing his work, Verschoyle said that it would be "of more interest to the wireless engineer than the aviator"! But the fact remains, with nothing but a comparatively crude apparatus, Verschoyle was pumping out packages of fairly high frequency waves, and the pulse rate was low—within the auditory range.

Since those early days, there has been, and still is, research conducted on similar lines all over the world, but how much of this is based on a Cosmic Matrix theory like the Theory of Unity is uncertain. Sometimes I have been amused to hear lay people, when knowledgeably pronouncing on antigravity, use terms of "one hundred and eighty degrees out of phase" and "either lagging or advancing with the currents of the universe," etc., etc., representing as they do further reasons for justifying the publication of this work. However, to be generous, despite the lack of applied logic, such remarks do have the right kind of label.

On-board Irradiating Power

When dealing with problems involving mass versus energy we are apt to be so mentally conditioned by fundamental scientific laws and conventions, that it is extremely difficult to attain a detached view, but we have to do exactly that when we are talking about lifting or accelerating a mass which, in a normal state, might weigh several tons, yet when gravitationally transparent becomes effectively massless; clearly the mass versus energy relationship takes on a dramatically different aspect.

Having seen how only a very modest force would be necessary to move such an otherwise massive body, the force can accordingly be expressed in normal electrical equivalents which, of course, will also be very small. This fact helps to put the situation in perspective when one hears of 'experts' pronouncing "that it would take the entire output of a small power station to move such a massive body at extreme acceleration"! Those experts are advised to cautiously rephrase their language when making such sweeping remarks.

Due in part to the nature of the process, together with the limitations imposed by the technology, with such a system there will be a tendency for a degree of cyclic instability, so that a body or vehicle loses weight through 'transparency' even to slight negative G, just like a 'soap bubble.' Therefore, the technology must incorporate an extremely sensitive auto-control system.

Bearing all this in mind, it will be appreciated just how much researchers are—or have been—penalised unless their work anticipates not only some kind of cosmic matrix, but one of an oscillatory nature. Remember that there are two fundamental ways in which to transfer energy from one point to another: by sending a *particle* from point A to point B, or by sending a *wave* from point A to point B, and of these the propagation of waves is the most efficient way to transfer energy. Apart from that, for the time being, the reader is asked to accept that by the application of an advanced version of the author's Syncomat process, specific parts of a vehicle—carrying its own small on-board system—are irradiated with pulsed packages of extremely high frequency EM wave energy to reach a gravitationally transparent state. There are of course many associated problems; we have only space for a few.

Orientation, Stability and Control

Some might suggest that rotation of parts of a gravitationally suspended vehicle would be necessary as a means of stability and orientation, while others would say that this could be more conveniently achieved by a solid state gyro system and so on. But it must be borne in mind that *unless suitably shielded or isolated, all* the physical contents of a gravitationally transparent vehicle would be void of inertia and therefore any 'normal' gyroscopic effects would be inoperative [Fig. 10.8 (a)]. Orientation of the craft would be a problem! Not only that, but the occupants of a transparent craft would experience weightlessness as they do with current spacecraft [Fig. 10.8 (b)]. Clearly, this would not be an overall perfect solution.

Now the foregoing cyclic tendency of plus or minus gravitational transparency can conveniently be reassessed as G1, G2 and G3 where G1 represents a marked reduction of weight, G2 equals zero gravity (gravitational transparency) and G3 represents an anti-gravity state—now to be described. All

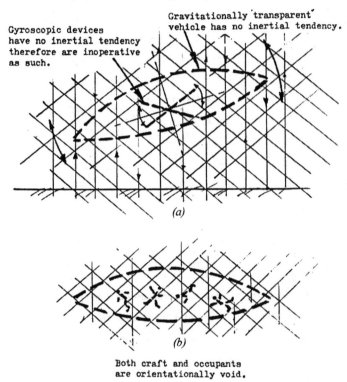

Gyroscopic devices have no inertial tendency therefore are inoperative as such.

Gravitationally 'transparent' vehicle has no inertial tendency.

(a)

(b)

Both craft and occupants are orientationally void.

Fig 10.8. In a totally gravitational 'transparent' vehicle absence of inertia prohibits mechanical stabilizing system and occupants remain weightless.

three states are fundamentally achieved by phase shifting in the methodology; in this case the Syncomat drive.

Consideration of these three states will reveal associated advantages and disadvantages depending on the operational intent. We have already dealt with the first two states; let us now turn to the last, but by no means least, of these.

The G Field Motivated Spaceship

During the past three decades or so, the author has seen many major advances in science and I am acutely aware of the changing attitudes in society.

Things which are taken for granted now were merely science fiction at the time when I first dreamed of rocket propulsion and journeys to the moon. Today the situation is not really very different, only now people accept rocket propelled space flight to the moon almost as a commonplace venture, yet my predicted journeys to our neighbouring planets in a matter of a week or so, is not even remotely taken seriously. So in order to help keep things in perspective, it somehow seems appropriate for me to yield to the idea of setting out the next logical projection in this account—which over the years I have had no reason to change—by reissuing this section from my original essay just as it was published in *1954 and 1965.*

Ever since the 1940s, I have passionately believed that once a way could be found to unbalance the implicit 'pressure' of the space matrix—possibly electrically—a vehicle would be accelerated just as if it were being moved by a second body as with gravity. It has since become evident this belief was not far wrong. Such a machine operating within the atmosphere at high speed—and therefore high power and extended field effects—would carry along with it a localised pocket of air. Due to the inverse square law however, this would be subject to a velocity gradient of diminishing intensity the further away from the craft it happened to be. Therefore such a machine would be more truly described as a spacecraft rather than an aeroplane.

Perhaps we might compare our present relationship with gravity to that of a small boy racing up a descending escalator; he has only to cease his efforts and he is promptly brought down again. Gravity, like the downward moving escalator, is being continuously expended.

Our future relationship to gravity may be more attractive and perhaps comparable to the same small boy suddenly discovering yet another escalator which will give him a free ride to the top, whereupon he changes back to the descending escalator to have an equally free ride down again. The whole point being that there is no comparison between his first frantically exhausting efforts to climb up and his leisurely stroll from one escalator to the other to achieve the same journey!

There is reason to consider our position is that of the small boy who has just begun to believe in the existence of yet another escalator and that if we pursue this belief, we too may find, not one, but an infinite number of such *gravitic* escalators (C ray modulations) which, we shall see later, may be manipulated by the expenditure of a comparatively small amount of energy.

Isolating Acceleration Forces

Modern astronautics has reached the stage when serious consideration should be given to the fact that gravity is but another natural phenomenon waiting to be explored, understood and put to useful purpose, as have light, sound, electricity and magnetism. From the aerospace point of view, one of the first observations we might make is the fact that gravity displays a unique and most advantageous factor in its make-up, that of uniform acceleration.

It is known that a body experiencing acceleration due to a gravitational field is under no strain whatsoever, for it is moving in a uniform field where each particle or atom forming its structure experiences a force equal to its neighbour. Therefore, there is an absence of *progressive* acceleration experienced by the particles. This condition is readily observed in the case of a falling egg [Fig. 11.8 (a)]. During the free fall, every particle of the egg experiences an equal force to its neighbour, therefore there is no relative movement between them. On coming in contact with the ground however, the eggshell is *decelerated* before the yolk, which still has momentum, therefore there exists a relative movement between them and of course collision with resulting structural failure as in (b). *Deceleration* in this case being identical to *acceleration*, i.e., *change* of velocity.

The point is more effectively illustrated by imagining that if we were falling towards the earth—that is, moving in the earth's gravitational field—we

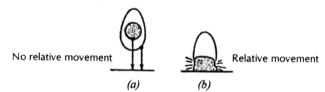

No relative movement   Relative movement

(a)   (b)

Fig 11.8. A falling egg offers a simple example of acceleration and deceleration, when all the molecules in the former are accelerated uniformly and decelerated progressively on impact.

would be *accelerating* at 32.2 ft. per second per second. As we have seen in Chapter 5, should the earth's mass then be suddenly increased to that of the dwarf star Sirius B, we would be subjected to a vastly increased acceleration, in fact something in the order of 25,000gs, but we would experience only a condition of weightlessness during the drastic change!

Now, if matter *is* a three-dimensional bond of carrier wave intermodulation and gravity is only a mutual unbalance of that bond, then, as stated, it should be possible to encourage this condition by other means. Once this is achieved, a spacecraft will move through space perpetually 'falling' with a minimum of energy expenditure—analogous to the little boy crossing from one escalator to another, or the sailor hoisting his sails.

It must be borne in mind that not only could this condition be produced in outer space, but also at the surface of the earth, it being possible to move a material object by this means in any chosen direction. Thus, should the 'unbalance' be created in the opposite sense to Earth's gravitational field, then a condition of so-called 'negative weight' would be obtained, the strength of which would depend only on the degree of unbalance, which of course is *true* 'anti-gravity.' It is worth noting here that a 'space drive' used to overcome gravity—be it rocket motor or any other *mechanical* contrivance—is *not* an anti-gravity device (popularly labelled by the press), any more than VTOL aircraft are.

Due to the nature of the field produced, a gravity wave motivated vehicle, together with its contents, could be moved at accelerations or decelerations of incredible magnitude. It would be capable of extreme manoeuvres, such as right-angled turns, or even being violently stopped or reversed; the effect would be exactly as 'falling' in any gravitational field and *the occupants would be completely unaware of any change.*[2]

On the other hand, the rocket—like an automobile suddenly moving forward—receives all its thrust via the motor and the astronaut is pushed rudely along by the seat of his spacesuit. In the case of the family car, first the wheels turn 'ejecting' a 'jet' of roadway. The resulting thrust is conveyed to the chassis microseconds later, and microseconds later still, the seat of the car is pushed also. Next it is the driver's turn and his body receives its dose of acceleration, but even so, the bloodstream, muscles and internals want to stay where they were before this rude interruption (inertia). The driver's nerve endings record this as an uncomfortable jerk and he is on his way. Usually this jerk is a fraction of a g, which may serve to illustrate what, say, 10gs sustained over several minutes is like!

On the other hand a useful approximation of what it's like being moved by a gravitational field, is that which most of us enjoy when tobogganing on a slope of, say, fifteen degrees or so, where the gravitational 'pull' component might be some fifty pounds, shared evenly by the structure of the sledge and our bodies, atom for atom. As we saw with the falling egg analogy, all the alternative step by step buildup of velocity—or loss of it—with its resulting discomfort and sometimes fatal consequences, would be absent with a vehicle moved by such a *uniform* field of force-like gravity. A glance at the sketch in Fig. 12.8 serves to illustrate the difference in principle further. The little railway trucks in (a)

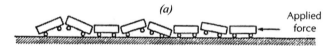

A progressively conveyed force producing structural failure as with cars, rockets etc.

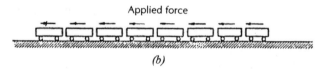

Fig 12.8. A uniformly applied force shared by the structure. By this simple principle there is no limit to which the structure may be accelerated or decelerated as in a gravitational field. This can best be illustrated slowly by a few ping pong balls on a suitable slide.

represent molecules in a structure. In this case, a force has been applied on the first truck giving it motion, but the other trucks stay where they were until the motion is conveyed by impact, truck by truck, or as the case may be, molecule by molecule. In other words there exists *relative* motion between them. But as in (b), should the original force be applied separately to each individual truck (or molecule) *at the same instant*, there will be a *uniform* motion shared by the trucks, and *no relative* motion between them. Such a condition might be imagined if all the trucks were made of iron being attracted by a powerful electromagnet.

With this principle, the velocities theoretically attainable would be colossal compared with any vehicle we have today, for as we have seen, the equivalent of the realisable Comat forces acting on matter might be measured in tens of tons per square inch, yet the energy input required to unbalance an extremely useful portion of this force will be relatively extremely small.

An End to Weightlessness

As is now well known, prolonged weightlessness, apart from being very uncomfortable for some people, can be a physiological hazard. It is interesting to note that space fiction writers conveniently—or perhaps intuitively—usually portray space situations in which the occupants of future, *non*-rotating spacecraft seem to be immune to weightlessness while in space transit. If it is mentioned in the dialogue at all, obscure reference is made to an 'artificial gravity.'

Extraordinarily enough, this is not so far from the truth, for once gravitational control is achieved, it will be possible to produce secondary fields

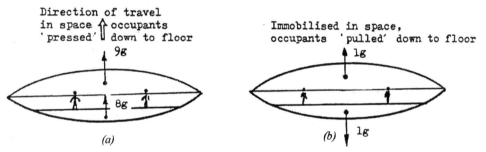

Fig 13.8. *Gravitationally propelled vehicle.*

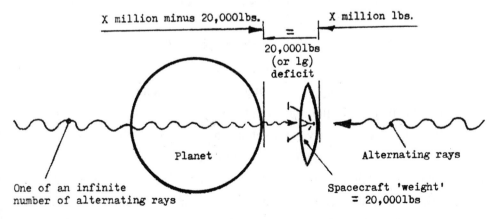

Fig 14.8. *Gravitational vehicle and occupants in normal gravity situation.*

to provide a normal earth environment of 1g differential between the crew and the ship. In a word it would be arranged for a spacecraft to be accelerated at, say, 9gs while the crew would be accelerated at 8gs [Fig. 13.8 (a) and (b)].

In 1954, the author devoted an entire chapter titled "The G Field Theory" in *Space, Gravity and the Flying Saucer* and later in 1964, an entire section on the interplay of G Field effects on vehicle and occupants.

Fig. 14.8 is merely a natural development of gravity as a deficit shown in Chapter 5 which, correlated with Fig. 15.8, should make the general principle clear. Also we have become so used to the conventional notion of rocket

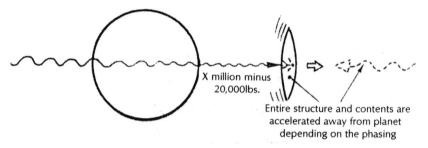

X million minus
20,000lbs.

Entire structure and contents are
accelerated away from planet
depending on the phasing

Gravitational vehicle and occupants in an 'anti-gravity' situation.

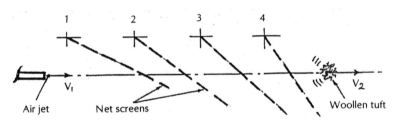

1   2   3   4

V_1 V_2

Air jet   Net screens   Woollen tuft

Air jet and screens analogy of the 'anti-gravity' situation where the residual air jet energy at V_2 is still capable of moving the woollen tuft away from the collected curtains. (See Fig 13.5.)

Fig 15.8. Illustrating the advantages of the gravitationally motivated vehicle

propelled winged reentry gliders like the space shuttle, that it is difficult for writers to visualise anything else. Though they have latched onto the idea that non-*reentry* deep space vehicles need not be aerodynamically refined. However introduction of the gravity wave motivated vehicle concept will introduce certain design changes. For instance, with this type of craft where weight may not be such a problem, it will be logical for the designers to incorporate generous floor areas, which combined with requirements for lift/propulsion and structural integrity and the above-mentioned 1g differential, inherently leads to the spherical or lenticular format, where the vehicle is propelled along its vertical axis both in VTOL *and* Space Transit. Should a more conventional flight mode be required while cruising *within* the atmosphere, separate multiple propulsion units can be employed, in which case at *low speeds* the lenticular shape would be aerodynamically more acceptable. A typical representation of such a craft is portrayed in Fig. 16.8.

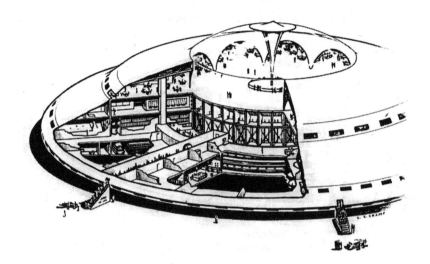

Fig 16.8. A large G.field motivated transporter.

A perhaps more mundane but nonetheless desirable advantage of a weightless free environment would be met at the social and domestic level where cooking and toiletry problems are entirely absent!

Predictable Composite Lift Effects

Of course there are many predictable physical side effects which would accompany vehicular interrelation with controllable G fields. As previously stated, I managed to fill a fair-sized book with some of them. If however we can list several here, which can be shown to bear some concordance with other, albeit unusual phenomena, then the proposition is somewhat reinforced.

Fig. 17.8 (a) shows a gravitationally 'transparent' vehicle hovering near to the surface—with arrows representing opposing gravitational waves in three directions only—together with an indication of the atmospheric lift component. Fig. 17.8 (b) represents a vehicle in the anti-gravity mode, i.e., in phase with the 'ascending' diluted gravitational waves. In which case the lateral thrust component could be imposed by the simple procedure of tilting as in (c). However, this technique might entail a possible hazard insofar as at high power rating and close proximity, some of the interface beneath the machine could become detached and rise with it, thereby—in some circumstances—causing a crater to be formed; which interestingly is an inclusion among many UFO reports [Fig. 17.8 (d)]. In *Piece for a Jigsaw*, when discussing the special circumstances of the formation of a crater, I pointed out that in many ways the

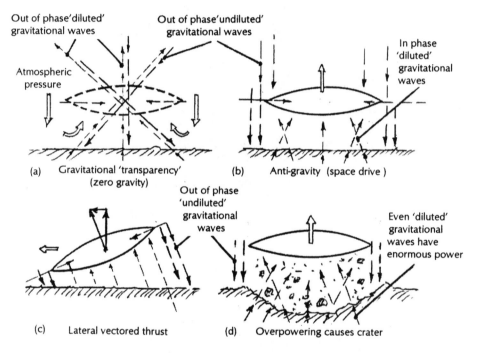

Fig 17.8. These predictable effects are supported by paranormal accounts
of levitation and UFO incidents.

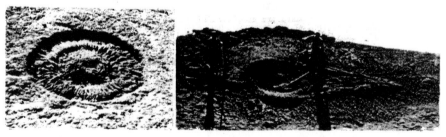

Fig 18.8. Crater formed in iron filings by a vertically ascending ring magnet.
Right, tapering crater formed by same ascending magnet moved laterally,
simulates effect caused by a vehicle taking off forwards.

effect a magnetic field has on iron filings was synonymous with the effect that a G field has on matter, the difference being one of degree, not of kind. I often illustrated this at lectures with the aid of a common powerful ring magnet and a few pounds of iron filings suitably coloured brown and green to represent grassland. Fig. 18.8 is a reproduction of the original plate from that earlier book in 1964.

It may also be useful to some readers to further simplify an anti-gravity state by correlating it with the jets of water and plate analogy shown in Fig. 19.8 (a). This shows a plan view of a plate supported on either side by a trolley on rails. As before, the effect of jets A and B cancel out, but if jets of water C are directed at right angles to the flow of jet B so as to intercept with it, the flow is diverted and prevented from contacting the plate [Fig. 19.8 (b)]. In effect, this represents a section through a 360[degree] domain, resulting in an unopposed movement of the trolley towards jet B. This somewhat basic analogy was

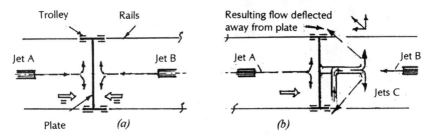

Fig 19.8. Water jet and plate analogy of the anti-gravity mode in which movement is obtained by lateral deflection of opposing jet. Note this is a purely hypothetical step analogy developed by the author for an aerodynamic concept.

originally first demonstrated in the 1950s in the family bathtub (to the amusement of all) when I was developing it for a lecture at London Caxton Hall and is similar in principle to well-known 'Magnetic Shielding.'

G Transparency Versus Vectored Thrust

It is also interesting to note that there are many recorded UFO reports in which an angle of tilt has been observed in association with longitudinal motion (helicopter style) and I devoted some time analysing this aspect in *Piece for a Jigsaw*. Even such basic investigation leaves little doubt that the G field operated vehicle would duplicate the process. How much reliance on such technological inclusions the sceptic is likely to adopt is conjectural, but the fact remains in the

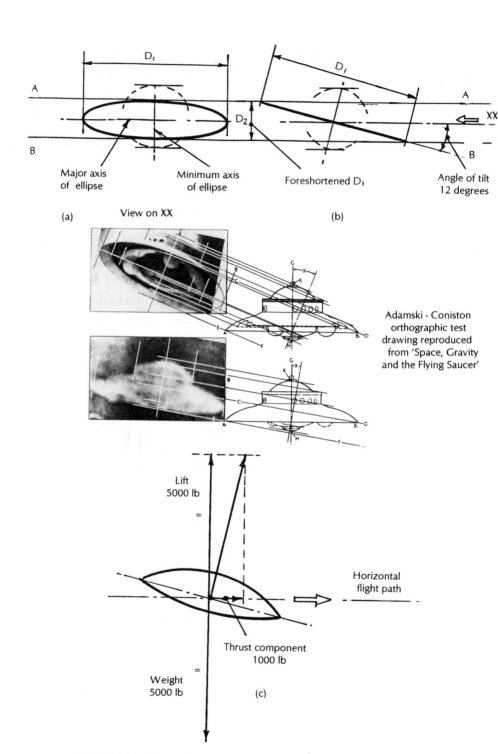

D₁

A ——————————— A

D₂

B ——————————— B

Major axis of ellipse   Minimum axis of ellipse

(a)   View on XX

D₁

A ——————————— A

XX

Foreshortened D₁   Angle of tilt 12 degrees

B

(b)

Adamski - Coniston orthographic test drawing reproduced from 'Space, Gravity and the Flying Saucer'

Lift 5000 lb

=

Horizontal flight path

Thrust component 1000 lb

=

Weight 5000 lb   (c)

Fig 20.8. Mutually corroborative support in a case of possible gravitational immunity determined by analysis of a UFO photograph.

present context it is useful to examine similar predictable phenomena.

For instance, claims have been made (and on occasion, supportive photographs have been taken) of such vehicles adopting the tilting posture while hovering with a complete absence of lateral displacement. Now the question arises, can the two behaviourisms not only be reconciled, but in addition offer mutual corroboration to the hypothesis in this book? The answer is an emphatic yes, as the following analysis will show.

In *Piece for a Jigsaw,* corroborative use was made of my orthographic projection analysis of the Coniston/Adamski photographs (reproduced here as an introduction to the process) which can quite accurately be employed, particularly with a photograph of an object having a circular planform.[3] Fig. 20.8 (a) represents such a hovering—but otherwise motionless—vehicle presenting a marked tilt towards an observer. From this it will be apparent that due to the fact that the object has a circular planform and the main diameter D_1 is at right angles to the observer's line of sight, it remains unforeshortened; on the other hand although D_2 is in reality equal to D_1, it *is* considerably foreshortened, therefore the observer records an ellipse. Now, by assuming the observer's line of sight passes somewhere near to the centre of the object, the situation can be laid out graphically to present a side view as in (b), where line AA and line BB are merely projections of the foreshortened diameter D_2. It will be apparent that if the unforeshortened dimension D_1 is transposed obliquely between lines AA and BB, a side view of the representative tilt angle is produced. Study of this will show that by this simple process, one is enabled to measure the corresponding tilt angle as it is presented to the observer, which in this case would be some 12 degrees as shown, and the situation can now be examined further.

Although as stated there are other photographs of UFOs in similar circumstances, beyond much doubt those reproduced in Ed Waters' book *UFOs: The Gulf Breeze Sightings* are among the best. One of these in particular shows the road over which the object is hovering, which offers an approximation of size, moreover it helps to access the observer's distance from the object. As an exercise, let us now imagine a similar scenario where the observer's distance is say 100 feet and a hovering vehicle having a diameter of approximately 18 feet, in which case its surface area would be approximately 254 square feet and although we cannot be certain that the vehicle might be extremely light or very heavy, as a first approximation we might assume the composition is similar to aviation standards of around 20 pounds to the square foot, which would give a weight of 5,000 pounds (or 2-1/2 short tons). So, for a *G field operated* machine to hover in that mode, it would have to have a vertical lift component of 5,000 pounds as shown in Fig. 20.8 (c). In which case it follows there would be a horizontal thrust component, tending to move the object towards the observer, of about 1,000 pounds.

Now acceleration $f = \dfrac{\text{force lb x g}}{\text{weight lb}}$ and substituting

We get $\dfrac{1000 \times 32.2}{5000} = 6.4$ ft sec.sec.

and maximum velocity Vm $= \sqrt{2 \times f \times S}$ distance in ft gives:

Vm $= \sqrt{2 \times 6.4 \times 100} = 35.78$ ft sec. approx mean velocity = 19 ft sec.

So that to cover 100 feet @ 19 feet per second would take the object a mere 5 seconds to reach the observer. In a word, any ensuing photograph would almost certainly be distorted, or at 1,000 pounds force component, the hopeful photographer would be lying flat on his back!

Now in this portrayal it has been allowed that the object is hovering *and* tilted, which implies one of two possibilities; the first being there is a secondary opposing counteractive force in operation (which due to the absence of visible local effects on trees, bushes, etc., rather beggars the question) or the vehicle was freely suspended in gravitation immunity, or 'transparent' state. All of which implies that if such cases can be attributed to hoaxes, then either the perpetrators are fairly technically qualified or they are chance favoured to an unacceptably high odds degree. But of no less importance, in the context of the main issue in this chapter, chance or otherwise, we see here evidence of a surprisingly supportive degree. Some may ask if an analysis of this kind would be complete without a meaningful contribution to one of the mysteries of our time.

"The Moving Finger Writes" . . . G Field Crop Circles?

Beyond the foregoing general discussion concerning the generation of such field effects, we are of course unable to discuss the actual physical process to any significant degree here, save perhaps to add that it is assumed, as with any other form of radiant energy, light, etc., it is possible to even beam a gravitational field.[4] In which case, all sorts of applications and scenarios spring to mind. For instance, as mentioned in Chapter 5 and elsewhere, there are reasons to suggest that in certain circumstances a localised field can produce rotary effects, which can be regarded as being similar in nature to the induced electromagnetic analogies also offered in Chapter 5, in which a 'descending' G field pulse might produce a clockwise effect, while its cessation would produce a reciprocal action.

Therefore, without undue effort, we can now imagine such a beamed pulsating field producing local rotary effects on nearby materials; in this instance, a field of corn. As the first pulse contacts the stalks of corn, they are affected uniformly over their entire length just as if they had become heavier, but due to a combined clockwise force they would tend to lay down in a tangential direction. Due to the fact that some of the stalks are lighter than others, the lay wouldn't be uniform. Fractionally, seconds later, the effect is withdrawn as the pulse ceases. Then the next pulse begins but this time the rotary force has been reversed to anti-clockwise and the next lighter layer of corn is effectively apparently woven until the process—occupying perhaps microseconds—is completed [Fig. 21.8 (a)].

Fig. 21.8 (b) is the natural development of the basic pattern from which we might conclude that a narrow coherent G field beam might be employed to produce, say, a ring outside the main circle or runs, etc. A useful clue would be if the ends of otherwise straight runs were radiused, which appears to be the case. Fig. 21.8 (c), (d), and (e) correlate some of the more well-known crop circle effects.

Of course I am not proclaiming how crop circles are created, but I am saying such phenomena is not at variance with the subject of this book and who knows but before long we shall see *triangular*-shaped 'circles.' But, for what it's worth, I do suggest the cause is more likely to be observed skyward and the world is being introduced to a cosmic symbolism.

The Exhaust of the System

With all prime movers, from the early steam locomotives to cars, aeroplanes and spacecraft, no matter how refined they may become, there are always accompanying losses incurred within the system. In this respect, an anti-

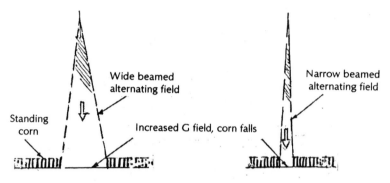

Wide beamed
alternating field

Standing
corn

Increased G field, corn falls

Narrow beamed
alternating field

(b) A G field can be beamed
much the same as light

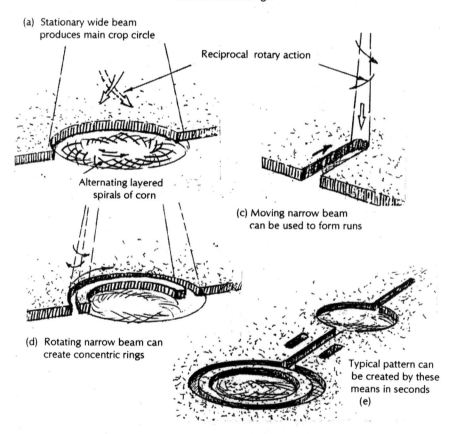

(a) Stationary wide beam
produces main crop circle

Reciprocal rotary action

Alternating layered
spirals of corn

(c) Moving narrow beam
can be used to form runs

(d) Rotating narrow beam can
create concentric rings

Typical pattern can
be created by these
means in seconds
(e)

*Fig 21.8. A variable focus pulsed G field beam could easily produce all
the otherwise mysterious crop circle effects. In much the same manner
as a rotary lawn mower does in grass.*

gravity technique of the kind we are envisioning will prove to be no exception, even though the proportional losses will be small. Verschoyle's quaint comment regarding this technology being more the province of the radio engineer rather than the aviator, has a bearing of truth, for we *are* dealing with radiant wave energy. Therefore the incurred losses will be in the form of spurious waves, which—bearing in mind the fact that we are manipulating probably the most dominant of all waves in the EM spectrum (gravity)—it wouldn't be surprising if these spurious waves were of a multifarious spectrum. These might include not only waves in the X-ray range, gamma ray photons, ultraviolet and large quanta of visible light, with all kinds of colours from ionised atmosphere and radioactivity (to name but a few), but towards the upper limits of our understanding, some have suggested there could even be apports and time dilation effects, all of which would vary according to the local conditions and mode of operation. Over the past four decades or so, I have been loosely describing this phenomenon as *the exhaust of the system*. Logically, much of this will be containable to an acceptable level, much the same as waste products are contained in all thermochemical processes. There is still another stage in the development of the Syncomat system, that of *electromagnetic* manipulation, which will not only be enormously powerful, but unbelievably environmentally 'clean.' In a word there won't be an exhaust at all! At which point it is as well to remember the words of Avenel, who said, "Some atoms of iron are arranged, or can be arranged, so that the extension modulations are not the same in all planes. This lack of symmetry can be encouraged by electrical means. It is quite possible that a single magnet removed from a powerful gravitational field would move through space of its own accord."

In echoing these words, it might also be well to point out that if to cause a body to be totally gravitationally 'transparent' (a convenient descriptive term I invented many years ago) causes the body to become inertialess, then we might loosely describe it as having *negative-mass*. Which is not without significance when it is remembered that this was the very term used by the visiting American General to my boss, Dick Jones,[5] when he said: "Damn it, man, give us a break, what are you getting for your negative-mass ratios?" Therefore we might be justified in asking, "If this desperate information bid was indicative of U.S. anti-gravity research involvement in 1956, what is the state of the art now?"

1 See Appendix.

2 The author emphasised this in *Space, Gravity and the Flying Saucer* in 1954 and Arthur C. Clarke independently arrived at the same conclusion.

3 Modern computerised graphics hadn't been invented when the author first introduced this application of the orthographic projection technique.

4 In *Piece for a Jigsaw* I discussed this in terms of a Point Source.

5 Chapter 1

9

ELECTROMAGNETIC PROPULSION
From Cars to Spaceships and Supermen

In setting out the following most important section we will continue to conform to the established method of examining the theory in allegorical terms, moreover we can do this with a high degree of confidence bearing in mind that mankind has been successfully exploiting electromagnetic phenomena at all technological levels (including space satellites, etc.) for many years without understanding what it *really* is, despite the fact that there has been written vast amounts of extremely complicated mathematical treatment analysis on the *effects* not the *cause*. Neither are we alone in this regard, for in his fascinating book *Engineer Through the Looking Glass* the brilliant original thinker Professor Eric Laithwaite voiced similar views when he said:

> There one gets the feeling that a voltmeter as an electromagnetic arrangement, is just another mysterious instrument that we do not understand. Apparently we are using it to translate the 'magnetic language' by the same shady means as those by which we first made the current in a coil produce an *entirely fictitious magnetic field*. To see is to believe they say, but to believe is not necessary to see.

Throughout this book the reader has been urged to consider that the phenomena of gravity, electric and magnetic 'attraction' and repulsion are similar and have a common cause. Based on this hypothesis, the conclusions arrived at in the preceding chapter indicate a technique for electrically reducing the 'opacity' in substances to induce gravitational 'transparency' on to an 'anti-gravity' state. Therefore, it would be quite logical for the reader to suspect that magnetism may similarly respond. Indeed, there are grounds to support that supposition, which has an added appeal when it is remembered that magnetic phenomena are comparatively more accessible for experimentation. Although available elsewhere, the following information may help to reinforce this approach.

In Chapter 3, I quoted a statement made by Michael Faraday in which he said, "I recently resumed the enquiry by experimenting in a most strict and searching manner, and have at last succeeded in *magnetising* and *electrifying* a *ray of light* and illuminating a line of *magnetic force*." To which I add a quotation from my book *Piece for a Jigsaw* in 1966, viz.:

> Under certain conditions, the plane of polarisation of a light beam can be rotated by the application of magnetic and electric fields. Three well-known effects are the Faraday effect and the Kerr Magneto-optic effect, in which light reflected from the pole of an electromagnet is polarised and the Kerr electro-optic effect, in which light traversing a dielectric medium, which is saturated by an electric field, has its plane of polarisation rotated. From which it is logical to assume that in some circumstances polarised light would have its plane of polarisation rotated by the strong associating magnetic and electric fields of a powerful G field.

The main point I wish to convey here is the obvious relation between photons and magnetic fields, which in this context are both regarded as extension modulations in the Comat spectrum. On these grounds alone, one might suspect that magnetic fields may be similarly manipulated by an adaptation of the Syncomat process and although any attempt at definitive methodology must be left aside for the present, there is no reason why we shouldn't examine some predictable effects here, which should be regarded as no more than an approximation. But first, another simple introductory analogy is appropriate.

Dynamic Electromagnetism

In Chapter 3, use was made of an analogy in which gravity and the ether were visualised in terms of an atmospheric pressure differential, and we saw how there is no reason why we shouldn't profitably employ the same analogy in the case of magnetism and the ether. We can now take this a step further.

Most of us are familiar with the old party trick which is also used by school science masters to demonstrate atmospheric pressure, in which a glass tumbler is filled with water and a piece of stout card is placed over the top. On being quickly inverted, people are amazed to see that rather than flooding out on to the floor, the card and water are held magically in place [Fig. 1.9 (a)].

This trick, although fascinating, is quite elementary and is of course determined by the fact that on the tumbler being quickly inverted, the water acts as an efficient piston which creates a partial vacuum or pocket of air at reduced pressure P_2, to be trapped between it and the bottom (now the top) of the tumbler. Thus there is a pressure differential created between this and the atmospheric pressure P_1 acting upwards on the card, thereby supporting the weight of the water. If not, the experimenter gets wet feet! It is, perhaps, fitting that such a simple demonstration should prove so vitally important as a step illustration here.

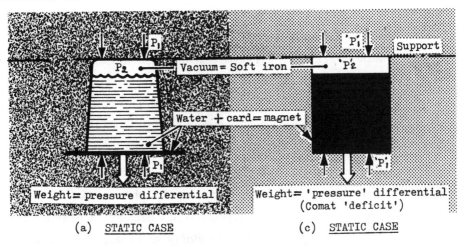

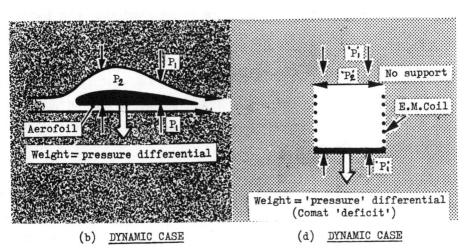

Fig 1.9. Hydrostatic versus fluid dynamic comparison analogy between permanent magnetism versus electromagnetic phenomena.

Now, this *static* pressure differential example is exactly comparable to the *dynamically* initiated pressure differential which is responsible for the lift on an aircraft's wings [Fig. 1.9 (b)]. In one case, we have a *static* pressure differential being contained within a vessel, which is employed to sustain a weight; in the other case, we have exactly the same effect created by a *moving* air stream.

To take the experiment on to the main purpose of the analogy, Fig. 1.9 (c) depicts an ordinary permanent magnet 'attracted' to a soft iron core. It will be seen that when this is considered from the Comat 'pressure' from without point of view, a very close resemblance emerges, in which the soft iron now represents the vacuum and the magnet represents the card and water.

Now this is very important, for according to Occam's Razor rule—i.e., a measure of common sense—it suggests that there should also be a *dynamic magnetic* example, which implies an *electrodynamic* case as in (d), which of course is the simple coil and iron core electromagnet. This is interesting, for we know that magnetic field strength is a function of ampere turns (current supplied times the number of turns of wire). Thus, for a given electromagnetic field strength, attraction or repulsion is directly related to current supplied.

However, it has been suggested that magnetic phenomenon is much the same as gravity and according to the hypothesis in Chapter 5, increased mass or 'density' produces greater gravitational 'shielding' and therefore greater 'field strength.' Appropriately, it is well known that magnetic substances are inclined to be the more dense ones, such as iron, nickel, cobalt, etc., which, to an extent, also supports the general hypothesis. But as stated, a very powerful magnet can be made from a heavy gauge coil of wire *carrying a high current*. Therefore, in this instance, we have an increase in effective 'shielding' or 'density,' produced *dynamically* due to a moving stream of electrons contained in the wire and *not due to an increase in mass*.

In the ordinary sense, the lay person might say, there is nothing extraordinary about that, but the phenomenon takes on a new significance when we remember to consider it in terms of the Theory of Unity, which in effect says that both the iron, coil, or the magnet—as Comat wave modulations—are tending to move past or away from each other in their travel back to source.

On the face of it, then, it appears that an electromagnetic Syncomat version *is* warrantable, but an in-depth investigation reveals the implications to be absolutely awesome! For when the enormous power of the so-called 'strong force,' as opposed to that of the 'weak force' (gravity, as shown in Chapter 3), is borne in mind, the adaptation of the foregoing descriptive term 'awesome' is seen to be more than justified. Indeed I am forcibly reminded of the fact that several decades ago, if devoid of a substantiating theory, someone had tried to persuade me that a simple energised electromagnet could be induced to move longitudinally or provide a thrust without the inducement of a close proximity paramagnetic mass, I would have been most incredulous. Therefore I must expect some readers to have similar reactions now and despite the logic in the arguments throughout this book, it nonetheless must again be reiterated: it is true or false *depending entirely on the efficacy of the Cosmic Matrix as postulated in the Theory of Unity.*

Before proceeding further, it is worth noting that from time to time one comes across apparent anomalies in nature which can eventually be explained. Magnetism has had its share, and as the reader is presently going to be confronted by an extraordinary proposition, the following preparatory example might prove useful.

The 'Magnetic Road'

I dare say, over the years, many of us have been tempted by the tantalising thought of using magnetism as a means of energy. As shown in Appendix 1, John Keely certainly did. It may be strange that this is so, for fundamental considerations deem it an impossibility, for in the *normal* sense, attraction and repulsion between magnets is in effect much the same as one can achieve with pre-tensioned springs. In other words, it is a two-bodied phenomenon. However, despite the logic of the apparently obvious, sometimes it pays to be just that little bit more curious, especially when we are confronted by demonstrable evidence which would indicate not so much that we are witnessing an anomaly, but rather a different application of the rules. The following offers a typical example.

Going by the reactions of visitors to my laboratory over many years, I realise this phenomenon is still largely unknown. The broad principle was initially presented to me by Antony Avenel in 1953, and my interpretation of it here was dictated by the proverbial lack of funds. Even so, it works and it has

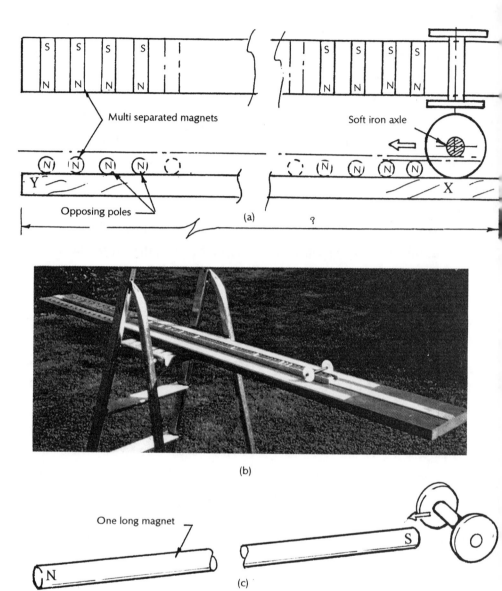

Multi separated magnets

Soft iron axle

Opposing poles

Y

X

(a)

(b)

One long magnet

N

S

(c)

Fig 2.9. The 'magnetic road'. This intriguing experiment is no more perpetual motion than a satellite in orbit is. Rather it is more of a logical sequence in the development of the general theory.

given me much fun and unquestionably helped along the way. As with other material in this book, to my knowledge this will be the first time it has been published.

It consists only of a pair of plastic wheels mounted on a thick soft iron axle, and several gross of permanent bar magnets, mounted on a board at carefully governed intervals, as in Fig. 2.9 (a) and (b). The diameter of the wheels must be such as to offer a working clearance between the magnets and axle when it is placed over them. The board must be perfectly flat and can theoretically be of unlimited length.

When the 'carriage' is placed at a determined distance from the first magnet—as in the sketch—and released, it is naturally attracted towards it. But then, if its acquired momentum is sufficient to carry it past this first magnet into the field of the next in line, the process is repeated and continued to the end of the track! In case some readers are inclined to think that the same thing would occur if only one long magnet of equal length is used as in (c), the answer is no, due to the extensive separation of the very *localised* magnetic poles. Neither should the device be confused with the clever domino stunts occasionally in vogue, for that phenomenon is more akin to the transference of *energy* by *waves,* whereas in the 'magnetic road' case, there is actual transference of a *mass* from X to Y, due only to an initial small 'thrust.' However, although it has step by step inference to the last developments of the theory, in this form the device is hardly of practical use, for it is difficult to imagine such a 'railway' festooned with miles and miles of magnets!

Some people have pointed out that in some respects this is reminiscent of Professor Laithwaite's linear motor railway, but we shouldn't be confused by this, moreover *this one* requires *no* power!

A Kind of Jet

It so happens that a natural, though not literal, development of the 'magnetic road' shows that in many respects a simple electrical solenoid can, in principle, be likened to a jet propulsion engine, in which the air normally 'sucked' into the forward end is replaced by a bar magnet and the axial flow compressor is represented by the energised coil [Fig. 3.9 (a)]. There is no fundamental difference between the two operations; the axial flow compressor creates a drop in pressure at the intake, which—as shown in Chapter 3—tends to move the engine towards the oncoming air (much as the *attraction* motion is shared by the coil and the magnet). If however, for the sake of the analogy, we assume that as the bar magnet enters the 'intake' of the solenoid, the coil is de-energised as in (b), the magnet—having momentum—will carry on through the

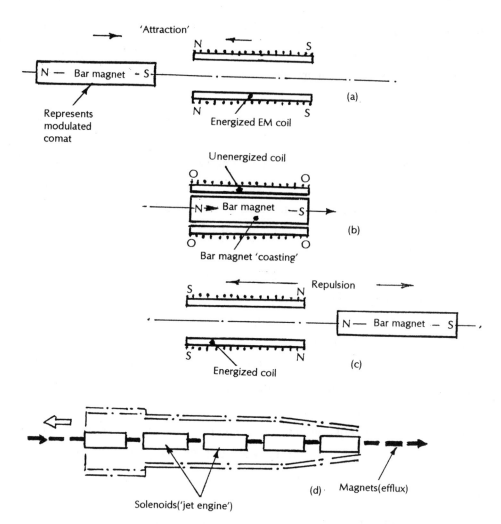

Fig 3.9. Step analogy as a kind of jet propulsion interpretation of EM syncomat drive.

unenergised coil until it begins to emerge as in (c), whereupon the coil is re-energised, but in reverse, so as to bring about a change in polarity which produces a *repulsion* between the coil and the magnet, in accordance with ordinary E. motors and the Newtonian law of action and reaction. From there, it is a simple step to imagine that if the bar magnets were very small and numerous, and provided the velocity and coil polarity changes were precisely synchronised, the small magnets would amount to a scaled-up string of molecules, or 'ejection mass,' being ingested into, and ejected out of, a kind of jet—or reaction—engine [Fig. 3.9 (d)]. At the risk of further complication, we don't have to stop there;

for example, we could assume that there were many other in-line coils and, in terms of power input, the 'thrust' would proportionally increase. This would be exactly analogous to increasing the compressor pressure staging in the turbojet.

We can now resume the main issue of the analogy by recalling that, according to the hypothesis, the physical structure of the bar magnets are nothing but an intricate conglomerate of modulated Comat waves being 'ejected,' or similarly synchronously interacting with another specially phased conglomerate of waves in the Comat, a process which I have loosely termed the EM Syncomat.

Single-bodied Electromagnetic Syncomat Locomotion?

In the natural progression from electrogravitic 'transparency' to gravitic levitation and propulsion, there is the fascinating prospect of an electromagnetic alternative, in which—with the exception of the energy input—it would seem there is very little difference in the technology. Initial interpretation might suggest that it would comprise a simple electromagnetic apparatus, which could be synchronously phased pulsed to provide an extremely high power-to-weight propulsion device, which would latch on to a monodirectional, relatively gargantuan, space ride! Later on, sober reflection will reveal that the solution isn't quite as simple as that. Nevertheless, this EM alternative would be light years ahead of what most of us could possibly have imagined.

Fig. 4.9 is a diagrammatic representation of a two-cycle EM Syncomat sequence, which can be studied in relation to the previous electrodynamic version. It will be apparent from this that not only would the EM version be experimentally more accessible and extremely more powerful, but apart from already well-known EM effects, there are no radiation problems involved. Moreover, it will be shown in the following epilogue that the EM drive carries an even greater significance.

Electromagnetic Power

A convenient and useful approximation of the degree of Comat imbalance—or power—which may be obtained electromagnetically, can be readily grasped by considering the performance of an ordinary industrial electromagnet, which is representative enough.

Typically, a unit such as that seen in Fig. 5.9 grasping a solid steel slab, with an average air gap of approximately 4 mm, can support no less than 14,500

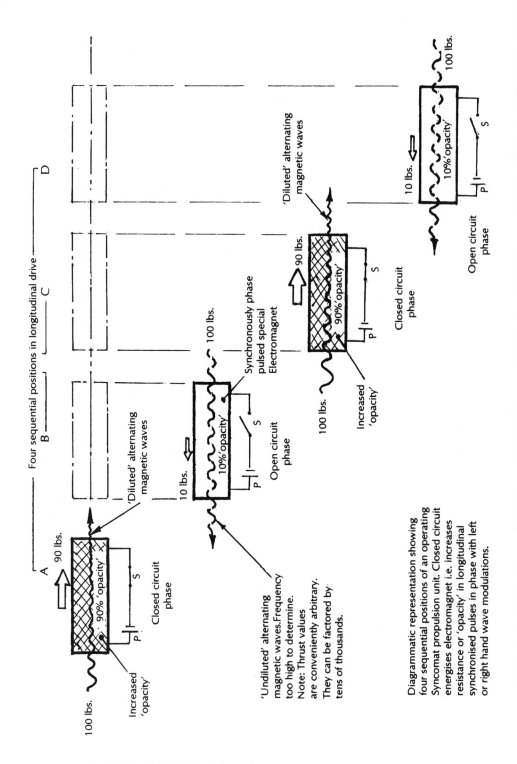

Fig 4.9. Variable EM 'drag' factor based on the author's 'syncomat' theory.

Kg (31,972 lbs., or 14.2 tons) for an input of only 5.5 Kw or 7.37 horsepower, which approximates to 1,967 Kg per hp (4,338 lbs.) or *1.94 tons per hp!* A more graphic evaluation can be obtained by comparing this with aircraft lift/propulsion systems. For instance, a high performance aircraft propeller might yield approximately 3-4 lbs. thrust per hp.; a ducted fan for lift purposes around 6-8 lbs. thrust per hp; and a helicopter rotor 8-14 lbs. lift per hp. This was one of the original chief motivations why this author has spent his life seeking a better way!

The above-mentioned air gap, although comparatively small, is an important feature in assessing electromagnetic power; for, as we have seen, magnetic phenomena also relate to the inverse square law. Therefore, with a

Fig 5.9. Even a typical scrap yard electromagnet yields extremely high 'attraction' forces for small power inputs.

Syncomat drive unit, the distances involved being microscopically small means that the magnetic 'pull' of the above industrial unit would be dramatically increased. An approximation of what this means in real terms is conveyed by assuming the air gap is reduced to 3 mm, in which case the 'pull' would become around 4-1/2 times that, say 19,500 lbs. hp. In the Syncomat unit, this force is in the form of a sequential half on, half off pulse, giving a force multiplied by time value. In other words the 19,500 lbs. represents a *constant* thrust of half that, i.e., approximately 10,000 lbs. hp. Again, the reader should remember I am proposing this force is *not* obtained from the magnet itself, but *directly* from the *surrounding space.* The principle is in no way limited to entropy, fuel in to work out efficiency ratios, or anything like our existing accepted interpretations of energy laws. It belongs to an *extended* order of things. In a word, the Syncomat drive situation is comparable to the sailor hoisting the sails of his yacht analogy offered at the very beginning of this book, no more, no less.

Size and Power Weight Ratio of EM Syncomat Propulsion Units

It must be reemphasised, for given electrical inputs, magnetic repulsion and attraction phenomena are of equal strength. But within the present context we are not talking about either of these, though the powers involved are of similar orders and therefore are representative enough for the present evaluation.

Magnetic field strength is measured in Gauss units which, in the instance of electromagnetism, is a function of core—or heel plate—size, coil windings and current input, etc. This information, together with designated air gap (to allow for inverse square effects) enables engineers to calculate the degree of magnetic attraction or repulsion.

Although the procedure is quite straightforward, such algebraic perambulations do not satisfactorily offer a graphic acquaintance for lay people. This can also be circumvented by further adoption of the previous scrap yard electromagnet example.

It so happens that the heel plate in that case has a diameter of 14 inches, giving an area of 154 inches^2, which (at the input horsepower of 7.37) together with the previously estimated 10,000 lbs. per hp, gives a magnetic field strength—or attraction—of some 73,700 lbs. over the 154 inches^2, equaling an attraction loading of around 478 lb. inch²!

To express this in even more graphic terms: if, for example, we consider a loaded family car of around 2,500 lb. weight and apply the aforementioned heel plate loading factor of 478 lb. inch², we arrive at a heel plate surface area of a mere 5.23 inches² or about *2.6 inches diameter*. Such a 'solenoid' coil, together with subsidiaries—cooling jacket, etc.—would have an overall length of 5.5 inches, overall diameter of approximately 3.5 inches and weigh a mere 6.5 lbs.,

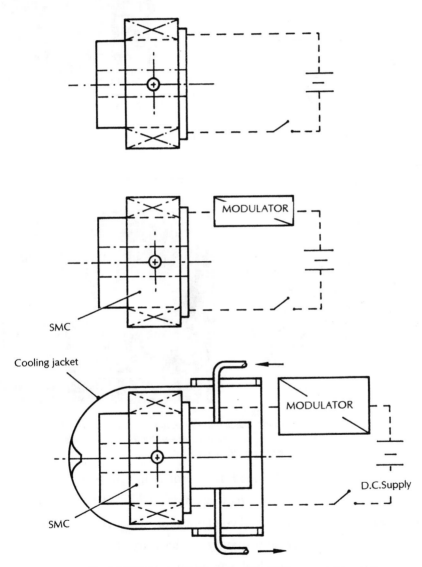

Fig 6.9. Schematic development of the EM syncomat drive unit.

giving an incredible power to weight ratio of 384 lb. per lb. At this same size and modest increase in power input, this little 'push engine' will raise the above car completely off the ground! Figs. 6.9 through 9.9 represent some contributions toward this end.

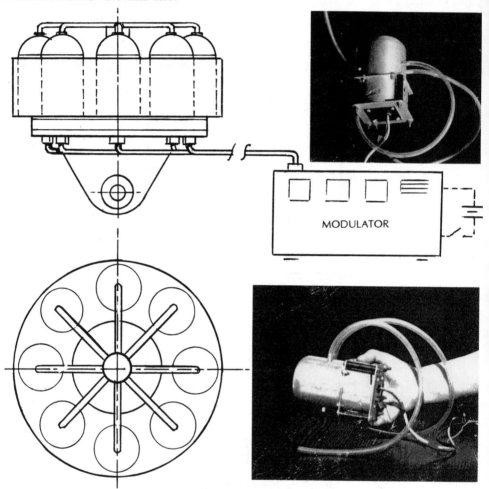

Fig 7.9. Syncomat impulse unit. The definitive version would easily propel a loaded family car up a 1 in 3 gradient. Several will lift the same vehicle completely off the rosd.

From this, it will be understood such propulsion units will represent a relatively small component of future vehicles, much as rocket motors–devoid of transmission gearing, etc.—are a small component of present day space boosters. On the other hand, so-called anti-gravity devices are symbiotic with radio transmitters; the radiation from which can either completely saturate the entire

Fig 8.9. Experimental EM syncomat impulse unit and support equipment.

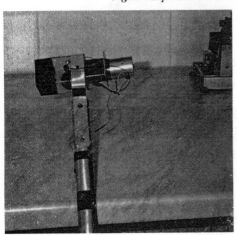

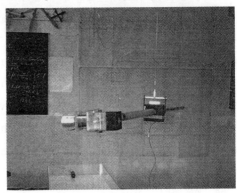

Fig 9.9. Experimental EM syncomat detector unit.

vehicle and its contents, or isolates them. It could be said that in this context the vehicle and its contents *becomes* a type of engine and people of a past age would look in vain for the 'prime mover' of such craft, for there would be nothing under 'the bonnet,' much the same as the UFO incident with which we closed the introduction to this book.

Not that there is anything unique about such a design, for we have been living in one since the day we were born . . . we call it a body, and furthermore we share the basic principle with all other living creatures, everywhere.

New Technology and New Specialists

In this context, it is interesting to contemplate that vehicle propulsion engineering—for all vehicles of the future—will be more the province of physicists and electrical engineers, rather than mechanical engineers and thermodynamicists. It will be noted that although the main proposition has been fairly well covered in analogous form, even from this it will be obvious that, given its authenticity, there are fundamentally but two chief factors involved. For instance, in this last alternative involving EM motivation, it is obvious there must be a suitably designed, rather complicated electromagnet, which has to take extremely high pulses (far in excess of commercially available E motors). Also, and by far the most important and difficult to research, the tuned modulator interface software. Although the technology for both these components is currently available, the mass usage would introduce a few other 'interesting' technological problems!

Among the more general queries often raised is, given the authenticity of the theory, wouldn't there be occasional evidence of coincidental frequency matching in all kinds of electrical machinery, particularly electric motors? Careful consideration will show that, particularly in the latter case, ordinary E magnetic pulses produced by such devices are far too coarse and unstable to 'latch on' to a whole range of harmonic frequencies, indeed such an interpretation would be tantamount to employing a vibratory road drill to locate a sustained radio signal.

As with any new concept, there are many associated problems to be overcome, and although the foregoing treatment as an analogous examination is as far as we can venture here, I have reasons to believe the fundamental hypothesis is correct. One of the most vexing aspects of this whole subject is the fact that, for the want of a wider basic theory, 'it' has been right under our noses all the time!

A Marriage of Convenience

This chapter would have remained incomplete but for the addition of an intriguing aspect, which I am sure will be recognised by many Ufologists. It concerns a very significant corroborative contribution, with which some readers will have become acquainted through the claims made by quite a few UFO contactees and, in particular, the late George Adamski.[1] It should be stressed that any similarity between their descriptions of purported vehicle interiors and the technological findings of the author are coincidental. In other words, being one hundred per cent devoted to the technical aspects of this work, I haven't the time nor the motivation to be solely directionally persuaded by such reports, that is, unless they happen to fit, which—as with many other aspects I dealt with in my books—they do. The following is a good example, but in order to view all the separate pieces of this latest 'coincidence' together, we must return to the three propositions discussed in Chapter 8, viz.:

1. The electrostatic space drive analogy (Type A)
2. Gravitational 'transparency' (Type B)
3. The G field motivated spaceship (Type C).

By now it will be apparent that the operation of all three of these is fundamentally similar, in that they employ a form of pulsed electric charges. In Type A, by means of the extension modulation phasing, 'attraction' or 'repulsion' motivation is generated. Whereas in Type B, the 'opacity' of the substances contained in the vehicle is completely neutralised to produce gravitational and inertial 'transparency.' In Type C, it is arranged for the particles in the substances contained in the vehicle to 'latch on' to the 180[degree] 'ascending' component of the alternating gravitational waves. So that, near to the Earth, the vehicle is virtually 'pushed' or 'falls' away from it and, in that sense, this is true 'anti-gravity.'

Each of these slightly different processes carries its own limitations and/or merits, and an advanced technology would take advantage of the interplay from all these vibratory fields for various applications. One application would be a shuttle vehicle which would operate close to a planet, and for certain rather technically complicated reasons, the choice would logically be the gravitational 'transparent' or neutral Type B.

It was shown in Chapter 8 that there would be no 'up' or 'down' for such a vehicle, and it and its contents would be weightless and inertialess, exactly like Father Dominic Carmo Dechaux in Chapter 7.[2] There would be no mechanistic method of orientating or propelling such a vehicle; by gyros and/or rockets, etc., for they too would be inertialess. This is a pity, for as we have seen, it would take the tiniest of rocket motors to hurl the energised machine at incredible accelerations, in which case both structure and the occupants would remain blissfully unaware of any change.

Format for a Spaceship or Vindication for a Scout?

In the section headed *Orientation, Stability and Control.* I said *"unless suitably shielded or isolated all* the physical contents of a 'neutralised' G vehicle would be void of inertia." But might it be possible to localise the effect, and are there any grounds for considering such a proposition? It so happens there is. The first example may sound obvious, yet it is valid. Common observation shows that there exists a kind of natural selection in nature, other than the biological kind. I refer to the apparent selectivity of magnetism. In other words, one can imagine a structure composed of varying materials such as aluminum, plastic and iron, which when immersed in a strong magnetic field, only the iron would become magnetised. Now that may be elementarily obvious enough, but when it is considered that all these substances—including the EM field to which they have been subjected—are in reality synonymous, then the situation takes on quite another significance, in that this phenomenon is a kind of selective 'screening' between nothing else but modulated *waves* in the matrix. I also have reasons to state that some of the articles of clothing worn by paranormally levitated subjects are not always similarly affected, in that they are sometimes seen to 'hang down.' Again, in the instance of the monk St. Joseph of Copertino, we notice that according to the testimonies of Surgeon Francesco Pierpaoli and Doctor Giacinto Carosi, St. Joseph was raised *above* his chair, which implies that either the chair was too heavy or physically secured, or the G transparency didn't extend significantly beyond his body.

If this field localising ability is accepted, it makes all kinds of situations possible. For instance, in certain circumstances, a section of a spacecraft could be subject to an anti-gravity acceleration of, say, 2g, while the occupants could be exposed to only 1g, giving them a normal 1g earth environment differential, as discussed earlier. Then, should a hovering mode be chosen (G transparency), the occupants could be excluded to remain at 1g. This is of paramount importance, because if it is possible to compartment the field effects, it *would* be possible to orientate and/or control an otherwise G transparent vehicle by means of rotating masses, preferably two counter-rotating gimble-mounted sections. For a given angular momentum, the larger these sections are, the slower they need turn, which makes for good engineering. By angularly offsetting the plane of rotation of these sections, the required controlling gyroscopic couples are imposed—remember, this is only possible due to the fact that these components are not G transparent. Next, such a vehicle will require only modest amounts of electrical power, therefore the two rotating masses can conveniently be 'doubled up' to act as an extremely efficient, high voltage generator of the Wimshurst type.

If you observed such a vehicle hovering at close quarters, first you would probably hear a low hum "like a swarm of bees" (which is typical of Wimshurst operations), you would possibly also notice a very strong smell of 'ozone' and if the machine tilted slightly you would notice two slowly counter-rotating concentric rings. If you looked closely, you would probably notice that coincident with the tilt, the plane of rotation of the rings was slightly displaced as in Fig. 10.9 (a).

The type of undercarriage or landing gear for such a craft would depend on the load-bearing requirements, and, from an engineering point of view, if the undercarriage is fixed, the number of ground contact points should be a minimum of three, because four points can be a nuisance if the vehicle is resting on an irregular surface as in (b). Also, for stability reasons, these contact positions should be fairly widely spaced towards the perimeter of the craft. Ground-handling bogie-type wheels would almost certainly be redundant, as such a vehicle could be quickly and easily moved around.

In Chapter 8, an electrical dictate for the shape of the vehicle was discussed, but there are other engineering considerations which have to be met. For instance, it is well known that the optimum formula for contained volume, surface area, strength, weight ratios, etc., is that which tends towards the spherical; nature demonstrates this fact with the egg. So, logically, the best shape for a space vehicle begins with the circular.

But there is more. Again, consider such a vehicle hovering in the G 'transparent' mode, in which the isolated crew have a normal 1g environment. In this situation, they could physically tilt the craft with the above gyro system, but

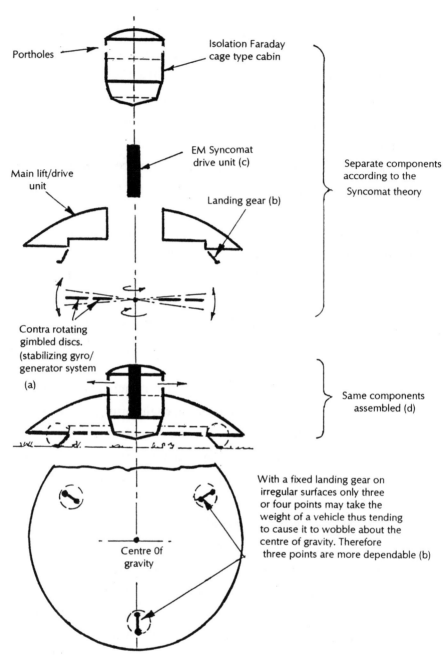

Porscholes

Isolation Faraday
cage type cabin

EM Syncomat
drive unit (c)

Main lift/drive
unit

Landing gear (b)

Separate components
according to the
Syncomat theory

Contra rotating
gimbled discs.
(stabilizing gyro/
generator system
(a)

Same components
assembled (d)

Centre Of
gravity

With a fixed landing gear on
irregular surfaces only three
or four points may take the
weight of a vehicle thus tending
to cause it to wobble about the
centre of gravity. Therefore
three points are more dependable (b)

*Fig 10.9. Assembled syncomat components for a G transparent craft produce a format
remarkably like the photographs taken by George Adamski. The profile of a gravity -
wave operated vehicle can be varied between those shown in Figs 16.8. & 10.9(e) or a
more prosiac format depending on the engineering requirements. It should also be
understood that with any of these variants from an aerodynamic point of view
external shape is of little consequence for the reasons I pointed out at
length in Piece for a Jigsaw.*

as we have seen, due to the transparent condition the vehicle wouldn't move laterally, there being an absence of a thrust component in that direction. This circumstance introduces the aforementioned 'marriage of convenience' situation, for this lack of thrust could be solved in no better way than by introducing the EM Syncomat drive system developed at the beginning of this chapter; moreover, the most logical geometrical positioning for this would be vertically—down through the centre of the craft—as in (c). Therefore, based almost entirely on the technological consideration in this book so far—without recourse or purposeful intent on this author's part—we have arrived at the format for our G field spaceship, which, coincidentally, happens to be identical to the archetypal 'Scout' configuration photographed by George Adamski [Fig. 10.9 (d)].

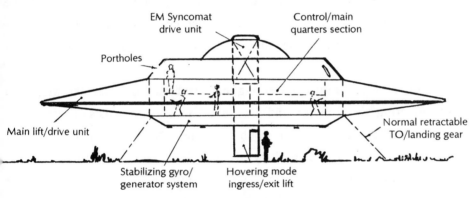

Fig 10.9(e) Showing configuration variant.

Concerning his description of the first close encounter with the scout ship in the desert in 1952, Adamski said, "As the ship started moving, I noticed two rings under the flange and a third around the centre disc. This inner ring and the outer one appeared to be revolving clockwise, while the ring between these two moved in a counterclockwise motion."

It is also interesting—and not without significance—to note that from his description of a later close encounter with the scout ship, near Los Angeles in 1953, Adamski said,

> I estimated the inside diameter of the cabin to be approximately eighteen feet. A pillar about two feet thick extended downward from the very top of the dome to the centre of the floor. Later I was told that this was the magnetic pole of the ship, by means of which they drew on Nature's forces for propulsion purposes, but they did not

explain how this was done.

And again concerning this encounter he said,

I noticed four cables which appeared to run through the floor lens (or immediately below it) joining the central pole in the form of a cross. Noting my change of interest my companion explained, 'Three of those cables carry power from the magnetic pole to the three balls under the ship which, as you have seen, are sometimes used as landing gear. These balls are hollow and, although they can be lowered for emergency landing and retracted when in flight, their most important purpose is as condensers for the static electricity sent to them from the magnetic pole. This power is present everywhere in the Universe. One of its natural but concentrated manifestations is seen displayed as lightning.

In the previous chapter, I included specifics to lend credence to both the theory and the claims of some UFO witnesses, and before we close this chapter I would like to add one further meaningful example, again with George Adamski in mind.

Continuing his narrative concerning the desert landing, Adamski said, "The ship was hovering above the ground" . . . and "Some of the gusts of wind were pretty strong and caused the ship to wobble at times." Now this is interesting, for even allowing for the fact that the craft may have been operating in the G 'transparent' mode, any aeronautical engineer will know that, even with our current technology, we have rate-sensors and backup techniques which can alleviate this magnitude of instability—even in model helicopters, leave alone an advanced, highly sophisticated spacecraft. Of course, the descriptive term 'wobble' immediately conjures up suspicions of gyroscopic perambulations, however we have already dealt with that. But there are other things (small to an engineer, significant to the observational ability of most lay UFO witnesses). However, first, in order to illustrate one other of these, a few basic sums are appropriate.

As an approximation, we can assume the estimated diameter of Adamski's scout was around 35 feet, giving a disc area of 962 ft.^{2}. Allowing the same density—loading as in the previous chapter—20 lb. ft.^{2}—we get a weight of around 19,242 lbs. (just over 8-1/2 tons). The profile cross-sectional area would be approximately 440 ft.^{2} with a drag coefficient of about 1.2, while a 'gusty' wind usually means something like 28 mph (40 ft. per sec.) over a period of maybe 3 seconds.

And according to fundamental laws of physics we have the following:

f = Acceleration ft sec.   F = Force lb.

g = 32.2   W = Weight lb.

d = Distance travelled ft.   P = Atmospheric density .00238

Cd = Drag coefficient 1.2   V = Velocity ft sec.

t = Time sec.   S = Projected surface area ft ^{2}

Now wind-drag acting on a body can be interpreted as a force (F) which can be found from the formula:

D = Drag lb. = $(.5\ P\ Cd\ SV^2)$ And substituting we get:

D = .5 x 1.2 x .00238 x 440 x 40 ft sec^2 = approx 1000 lb. Also:

$f = \dfrac{Fg}{W} = \dfrac{(1000 \times g)}{19242} = 1.67$ ft sec per sec. And:

d = $.5\,f\,t^2$ = .5 x 1.67 x 3^2 = approx 7.5 feet (scout ship sideslip)

Although this would be the case if the vehicle was not gravitationally 'transparent'—weight equals 19,242 lbs.—we can assume this situation due to the fact that, in the hover mode, allowance must be made for Archimedian buoyancy lift. So the G transparency factor would be considerably subdued. This, together with the uncertainty of the actual weight, means the above 7-1/2 ft. displacement can only be representative. But the whole point is, this represents a visual measure of what could be taking place. In other words, the 'marriage of convenience' format previously set out would provide the logical way to offset sideways drift, that is, by tilting the whole vehicle *into* wind, helicopter-style. Is this just *another* coincidence? I, for one, ran out of argumentative excuses long ago!

It must again be emphasised, in offering this book, I haven't set myself the specific task of helping to redeem someone I am inclined to believe is a much maligned man; but nevertheless I welcome this opportunity to do so.

Design for a Deep-space Carrier (Mothercraft)

In the introduction to this book, I referred to closing remarks I made in *Piece for a Jigsaw* with regard to the type of UFO designated 'Mothercraft.' I now welcome this opportunity to touch on this aspect here. However, it should be understood that the following speculation is based chiefly on what has been said so far in this and earlier chapters. I am well aware of the fact that the ultimate development of the previously described 'scout,' along with gigantic deep-space carriers, is as far from this portrayal as are paddle steamers from the QE2. Nevertheless, it is interesting to see how representative one might get through the logical application of established scientific principles as follows.

Consider that it is required to design the structural layout of such a massive deep-space carrier so that—motivated by the foregoing process—it will be able

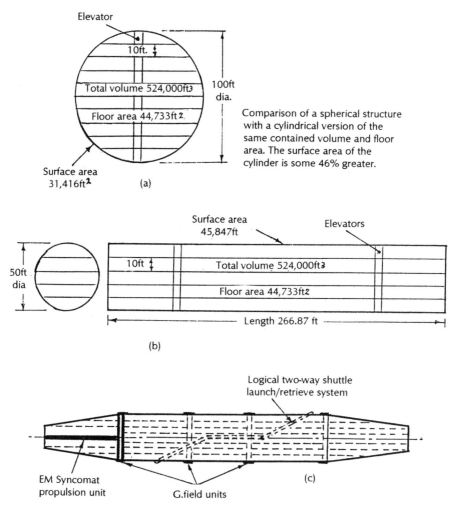

Comparison of a spherical structure with a cylindrical version of the same contained volume and floor area. The surface area of the cylinder is some 46% greater.

Fig 11.9. The logical development format of a large deep-space carrier in terms of the syncomat theory.

to operate within the atmosphere of a planet as well as in space. In *Piece for a Jigsaw*, I showed how such G field operated craft are not subject to aerodynamic drag forces, thus refinement in shape is not a prerequisite, which considerably reduces the structural requirements. In other words, all we need to consider here are the parameters for a basic box structure as a first approximation.

With the emergence of airships, engineers soon became aware of a natural principle in nature which was duly referred to as the 'area-cube law.' It, in effect, shows that various shapes of bodies offer better—or worse—ratios for contained volume versus enveloping surface area. Nature adequately demonstrates this truth in the instance of a hollow sphere, which has the greatest contained volume for the minimum surface area. Two of the important consequences of this rule lie in the fact that surface areas of a structure largely determine its weight, and the sphere is inherently the strongest shape [Fig. 11.9 (a)].

This doesn't necessarily mean that all space vehicles should be basically spherical, there are other factors involved, but it does indicate that designs should be influenced by that structural derivative. Fundamentally, the closest alternative compromise is, of course, the cylindrical—as is evidenced by the designs for modern orbiting space stations—in which the surface area versus volume and weight is not so efficient, but better from structural and utility considerations [Fig. 11.9 (b)]. A more obvious advantage of the tubular configuration is, of course, the interconnecting elevator height, which is halved. From a purely structural stiffening point of view, it is logical to form the extreme ends of large tubular bodies towards the conical, streamlining being unnecessary.

Fig. 11.9 (c) carries the observation to its conclusion, in which EM Syncomat propulsion and G field units are arranged at the aft end and the middle of the vessel, so that in terms of the reasoning offered here, we arrive at a perfectly lucid format for a deep-space craft, which many UFO readers will recognise as the familiar Adamski Mothership.

Interplanetary Journey by G Syncomat Drive

Many people ask, "What effects would all this have on what is, after all, one of the most significant aspects of a future space drive technology, specifically time and energy requirements for a given journey? Can this be related to the

Syncomat proposition?" Bearing in mind that in the anti-gravity mode there will be no restriction to acceleration, together with the extraordinary efficiency of the technology to be employed, as a first approximation we can relate to the information on energy requirements set out in Chapter 3. This illustrated the three main levels of energy conversion: the thermochemical, the thermonuclear and the theoretical direct conversion of mass into energy. In order to convey a visual conception of what this means in terms of space flight, we can choose an example based on the last of these.

Consider the case for a hypothetical lunar spaceship of 30 tons weight. For simplicity, we assume that the vehicle will be accelerated to the moon at constant acceleration for half the journey, then decelerated at the same value for the remaining half, though this need not necessarily be so for this type of spaceship. For instance, depending on the type of mission, the crew may decide to accelerate at a constant 2g for the major part of the trip, and then decelerate at, say, 10g for the remainder. In any event, neither they nor the vehicle would experience the least disrupting stresses. For this particular exercise, however, we shall neglect the accelerations of the earth-moon system and suppose the craft to be operating in a constant 1g acceleration/deceleration field over the entire journey and set this out as follows:

Distance of the moon	=	238,857 miles
Half distance of the moon	=	119,428 miles
Acceleration of ship	=	1g (32.2 ft. per sec. per sec.)
Weight of ship	=	30 tons (67,200 lbs.).

Now according to the laws of motion we have:

$$V = \sqrt{2fS}$$

where V = Final velocity in feet per second.

f = Acceleration, i.e., 32.2

S = Distance in feet

W = Weight

t = Time

and substituting, we have:

$$V = \sqrt{2 \times 32.2 \times 119,428 \times 5,280} = 201,800 \text{ ft sec.}$$

But V also $= ft$

$$\therefore t = \frac{V}{f} = \frac{201,800}{32.2} = 6,250 \text{ sec.} = 1.735 \text{ hr.}$$

Now 1 hp. hour is equal to 550 foot pounds of work times 3,600 sec., or 1.980×10^6 ft lb.

But the work done = WS = 67,200 x 119,428 x 5,280 ft. lb.

$$\therefore \text{H.P. hr. expended} = \frac{67,200 \times 119,428 \times 5,280}{1,980 \times 10^6} = 21.4 \times 10^6$$

And 1 kW is equivalent to 1.34 h.p. Therefore we can write :

$$\frac{21.4 \times 10^6}{1.34} = 16 \times 10^6 \text{ kWhr.}$$

We know that one gram of matter, totally converted into energy, could yield 25 million kW for one hour. Which means the mass consumed in the liberation of 16×10^6 kW/hr. would be $\frac{16 \times 10^6}{25 \times 10^6} = 0.64$ grams!

But this calculation was for only halfway to the moon, and because it will take just as much energy to decelerate the machine at 1g as it does to accelerate it at 1g, we can assume the mass consumed to be exactly twice, i.e., 1.28 grams. Which means, if totally converted, there is sufficient energy in an ordinary mass of 2.56 grams to send a thirty-ton spaceship to the moon *and back* in a little under seven hours!

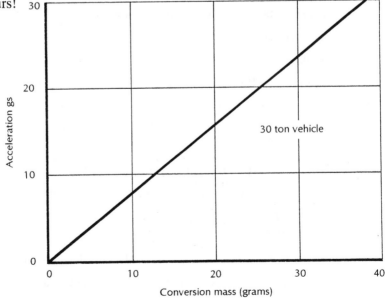

Fig 12.9. Mass/acceleration factor for a thirty ton lunar space craft.

The acceleration-mass conversion graph in Fig. 12.9 shows the straight line relationship. From it, the following startling facts emerge.

It would take just over one hour to send the same vehicle to the moon at a constant acceleration and halfway deceleration of 10g, for the total conversion of approximately 12.75 grams of matter; .78 hour for the same ship, accelerating at 20g, at a consumption of 25.6 grams; while, if we like to be rash and expend the exorbitant amount of 76.8 grams, we could send the ship to the moon and *back* at a comfortable 30g in a little over one and a quarter hours (or to be precise, 1.272 hours). In other words, during the trip, it is quite possible that the crew might drink considerably more fluid than the 2 to 3 oz. conversion mass required by the ship to accomplish the entire journey!

Perhaps even more fantastic by modern rocket technological standards, the same 30-ton ship could be accelerated to the planet Mars at its nearest (35,000,000 miles), at a constant acceleration/deceleration of 50g for the dematerialisation of approximately 20.5 pounds of mass, and would take a little under six and a half hours to do it! This means, that at the halfway point, the ship's velocity would be roughly 3,270 miles per second, but even this is only 1.76% of the velocity of light.

In the foregoing examples, for simplicity, no allowance has been made for the previously discussed 'exhaust' of the system; neither has the mutual gravitational effect between Earth, moon and Mars systems been included.

Although the general Syncomat principle possesses an extremely efficient alternative space flight technology, it must be understood that it eventually involves an electrodynamic process, which occupies a spot between the thermochemical and the thermonuclear levels. The fact that a Syncomat spacecraft of the type we are considering would be both inertialess and acceleration free (spread over a correspondingly short time) places the *actual* fuel expenditure versus time ratio above the upper limits of the thermonuclear level shown in Chapter 3. Using the lunar and Martian forecasts, we can make the following rudimentary assessments.

A 30-ton G Syncomat spacecraft could travel to the moon and back at 30g in approximately seventy-five minutes for the expenditure of around 24 lbs. of fuel (3-1/2 gals.).

The same 30-ton spacecraft could travel to the planet Mars and back at 50g in approximately seventeen hours for the expenditure of 5,120 lbs. of fuel (731 gals.)! If this sounds too incredible, we must try to remember that barely a lifetime ago, a limit was imposed on the number of seats and a maximum speed of 20 miles per hour set on 'horseless carriages' of that time. The descendants of these are so prolific today that certain types of *children* steal them and cause mayhem at over 100 miles per hour in our streets and towns at night. The future should be . . . . interesting !

For the present, we must leave the anti-gravity Syncomat spaceship to that future and devote the remaining pages of this chapter to an application of the Syncomat principle, which is much closer to actuality than most would ever dream.

The EM Syncomat as a Universal Drive System

By now the reader will have become acquainted with the fact that, despite the unquestionable superiority of gravity as a space drive, the associated technological difficulties are formidable to say the least; we have reviewed just a few of them. On the other hand, it has been shown that the associated problems with electromagnetism are numerically few. Therefore, a natural progression in development suggests that EM Syncomat should accompany or even precede gravitation research; that is a true evaluation of the present situation. With that in mind, we can continue by first examining the EM version in terms of space flight. However, due to the fact that this method is more restrictive there are other conditions imposed which require more detailed definition.

Although solely electromagnetically propelled craft would not be isolated from acceleration and deceleration inconveniences as would gravitationally propelled vehicles, they would nevertheless be vastly superior to the rocket techniques in use today. As we have seen, within the atmosphere, lenticularly shaped spacecraft would fly longitudinally, presenting the most efficient shape to the airflow [Fig. 13.9 (a)]. However, in airless space, such a craft would accelerate at constant 1g towards the halfway mark orientated obliquely, or nearly so, in transit to its destination, as in (b). At the halfway point, it would cease the acceleration (power shut down) and the craft, now coasting, would rotate about its central axis to face the other way, during which manoeuvre the occupants *would* become weightless, unless secured within their seats [Fig. 13.9 (c)].

Upon completion of rotation, the craft would then employ its EM Syncomat power units at 1g to absorb, or brake, the enormous kinetic energy it possesses, thereby restoring normal earth environment for the remainder of the journey. Thus, one has to get used to the idea of approaching another planet by *visually* flying *up* to it until the halfway mark, then after the orientation manoeuvre, descending in the normal manner.

By this means, the time taken for a return trip to Mars at the constant halfway acceleration-deceleration of 1g would be approximately 3-5 days. It is interesting to compare this with the recent American proposal to send a rocket to Mars, which will require the crew to remain in a weightless condition for *over 18 months!*

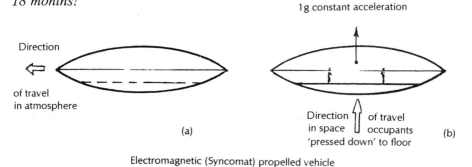

Fig 13.9. *The most logical method of providing an earth type 1g environment is to take advantage of the constant 1g vehicular acceleration/deceleration technique and orientating the occupants normal to the acceleration.*

An interesting tradeoff to the new space flight technique lies in the fact that due to the reduced interplanetary transit times, the orbital position of Mars after only three and a half days would *not* have changed very much, allowing a practically direct visual approach, with very little course correction. This would drastically reduce the otherwise complicated orbital phasing technique required.

Because of the wide divergence in the scale of masses and size involved, it is tedious to quote meaningful data. However, based on the previously quoted thrust figure of 10,000 lbs. per hp., a thirty-ton EM Syncomat Martian vehicle would require an unbelievably mere seven or eight horsepower to initiate and sustain the operating cycle, and from then on it can be assumed the machine

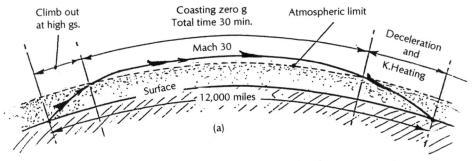

Flight pattern of a Trans-atmospheric Rocket Plane

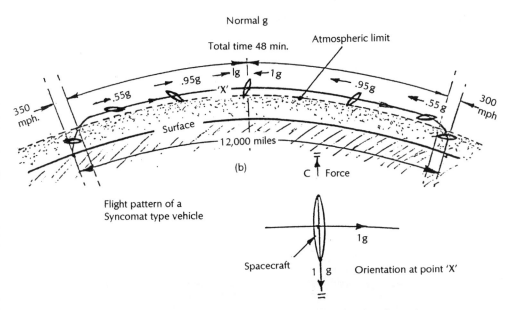

Fig 14.9. Comparison of the flight patterns for a projected rocket plane of the late 1990's compared with a similar mission of a syncomat propelled vehicle. Although the flight time would be marginally longer, it would generally be a far more comfortable and economic trip by this method.

would operate on solar power in space.

By comparison, even a brief enquiry reveals that the extraordinarily vast sums being spent on aerospace research include investigation into Trans-Atmospheric vehicles, which would have speeds around Mach 20 and 30! This, of course, would require the crew and passengers to be similarly subject to weightlessness for most of the time, which could be about one hour in some instances. Then, on the descent leg of the trip, the stored kinetic energy in the vehicle has to be dissipated, either by retro-fire and/or by atmospheric burn-off, with all the pyrotechnic problems that that implies.

Fig. 14.9 (a) compares such a trip by rocket plane with a VTOL Syncomat propelled vehicle employing the same halfway acceleration/deceleration as before (b); but in this instance, the vehicle is accelerated forward not by 1g as before, but an average of .8gs. This, together with diminished earth gravity and orbital centrifugal force, would produce earth conditions, allowing passengers to walk about. Ascent to above atmosphere would be at about 50 degrees, lasting approximately 5 minutes on the upward and downward leg, at maximum speeds of little more than 350 miles per hour, thus completely negating existing atmospheric kinetic heating problems. This attractive method will enable vehicles to be constructed more on existing aircraft standards, rather than those of high velocity ballistic missiles! [Fig. 15.9]

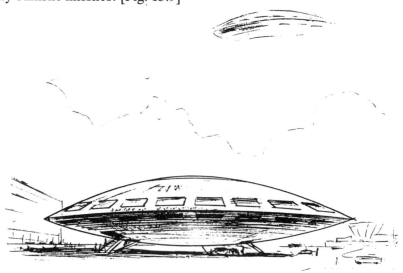

Fig 15.9. Large cruiser class trans-atmospheric syncomat vehicle.

A New Superman?

Now should the reader have been inclined to find the author guilty of peppering this work with somewhat disparaging misgivings, I welcome this opportunity to redress the balance a little by lightening this peep into the future with a more amusing—though with no less serious intent—example of Syncomat things to come.

In Chapter 7, the reader's attention was drawn to the fact that, in terms of antigravity, the aerial flying antics of 'Superman' might not be so far out after all. The origin of this style of travel might have been inspired by the experimental 'retro-packs' strapped to the shoulders of willing participants—including 007 James Bond!

Initially, these were either liquid fuel rockets or very small turbojet engines, in all weighing quite a few pounds and not always reliable. No doubt at this stage the reader is already way ahead of me, so all that remains are a few specifics, thus:

The average man weighs about 161 lbs., however, for this particular exercise, we can afford to be generous and call it 168 lbs. and even give this a times 2 safety factor of, say, 336 lbs. In terms of the previous thrust rating of 10,000 lbs. per horsepower, 336 lbs. would require only .034 hp (or a mere 25 watts) which, at 12 volts, gives about 2 amps. For this approximation we can assume the power loading related to the magnet core plate area is approximately 324 lbs. in.^{2} which, for the above-required loading of 336 lbs., gives just 1 in.^{2}.

Now in the case of the above retro-packs, designers had little option other than securing the unit at the back of the wearer, which incurred a forward centre of gravity problem. By comparison, a small Syncomat unit—as in the case of a similarly powered car—can be divided into, say, 8 separate radially disposed small units around the wearer's centre of gravity, which happens to be the waist.

These small Syncomat impulse units, together with an integral rate sensor/control monitor for the three axis control (pitch, roll and yaw), could be separately or collectively controlled and would measure approximately one eighth of an inch thick, just under half an inch wide and a little over four inches deep. These, together with in-between power packs, could be conveniently woven into a deep waistband as part of a specialised sky-type suit, while the operational

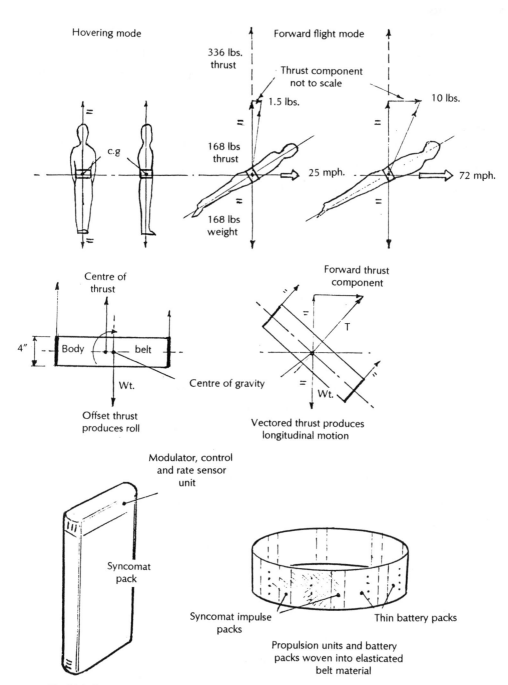

Fig 16.9. Syncomat survival suit belt, aerospace or underwater mobile. This might seem to be a long way from the gravitational transparency excursions of people like Father Dominic Carmo but it is a step in the right direction.

duration would be about one hour. Thus we have arrived at a futuristic version of the ordinary parachute [Fig. 16.9] and a few interesting 'performance' specifics will not be amiss.

The best aerodynamic lift over drag ratio (L/D) for the average person is around .55, where the coefficient of lift (C_L) is about .35, while the coefficient of drag (C_D) approaches .63.

At the generous loading we have quoted, i.e., 336 lbs., we can assume a vectored forward thrust component of, say, 1.5 lbs., which would produce over thirty feet per second or 25 mph forward speed. While a more generous 10 lb. vectored thrust would produce a useful 72 mph!

Although we can assume the 'pilot' could employ a simple weight shift technique for control—similar to hang gliders—and/or manual selective or collective Syncomat adjustment, it can be anticipated that the direct visual 'feedback' control currently being developed could be used with advantage. *Given only* the technology portrayed in this book—for those so inclined—let there be absolutely no doubt about it, all other things being as they should, people *will* indeed be able to fly!

To some lay readers this will seem like science fiction; to most aerospace engineers—given Syncomat availability—quite straightforward; and to specialists of a future culture, it will be politely dismissed as being rather crude.[3]

Of course, there would be countless other uses for such a device, particularly in the role of recuperation and aftercare for the paraplegics throughout the world.

In the first chapters of this book, the reader was introduced to a little-known theory, and throughout the remaining chapters I have offered evidence for that theory, together with my own interpretations of it. As an epilogue to the last few chapters, there has been included some of the more implicit ramifications and more obvious examples of a totally encompassing new technology. However, there is one last step to take in this account of the Syncomat story, and with it we shall see how amazingly such a system might be self-perpetuating—*together with all that that implies!*

1 *Flying Saucers Have Landed* and *Inside the Space Ships.*

2 There are many UFO sightings describing apparently inverted or standing
  on edge hovering discoidal vehicles.

3 In fact, this scenario will not be unfamiliar to the many Ufologists around the world.

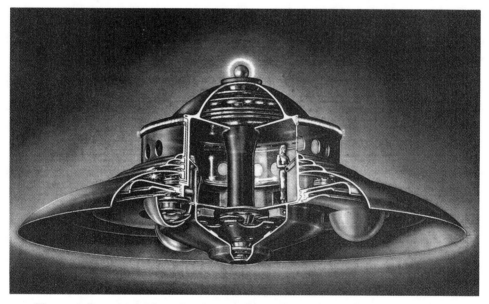

The original painting by the author depicting his impression of the interior of an Adamski
type scout ship shown in the book 'Space, Gravity & the Flying Saucer' in 1954.

The author's mother identified this craft as being like that which landed 'brightly lit' in front of
our country home in August 1961 during the CE3 Ufo episode mentioned in the introduction to
this book and fully described in the AT FACTOR. Concerning which, several high ranking
Air Force friends privy to my lifetime research in this field, suggested the literal translation
of the RAF motto was commendably more appropriate i.e. 'Through adversity to the stars'!

EPILOGUE

Not Least Unlimited Clean Free Energy

If, for some, the logical development of Syncomat propulsion set out in the last chapter is apt to stretch credulity to the limit, the next logical step projection may be even more difficult to accept, for it requires a change of habitual thinking, which some might say is of 'quantum leap' proportions. Such a reaction is natural when we bear in mind that most people are so conditioned to accept existing values of the time, that a totally different way of looking at things is difficult in the extreme.

For example, in retrospect, most of us can experience a measure of such reaction by spending a few hours at a museum and getting acquainted with a comparative mammoth-sized beam steam engine of the eighteenth century. Typically, it could have a weight of over a ton, turn at about 8 revolutions per minute, generate a few horsepower and would have taken many months to build. Today, the I.C. engines which I used in many of the test models shown in my book *The A.T. Factor,* turn at over 20,000 rpm, weigh under 3 pounds and can fit into an average size pocket. Moreover, their build time is only a few hours.[1]

Therefore, in presenting this last projection, in essence I am asking the reader to take that comparison exercise a step—granted a very large one—further, by imagining those little aero-engines becoming smaller than they are today. But instead of an accompanying change from coal burning and steam to nitro fuel burning, we have an arrangement of small special EM coils which require an incredibly small amount of energy input for a very high gain. So far, this much readers may have been able to accept, but now it has to be said in so doing we are, by implication, staring so-called perpetual motion in the face! Let us now see why that is probably so.

A typical dictionary definition of the term 'perpetual motion' is stated as "a machine which would go on working forever without receiving energy from an outside source," which clearly is *mechanically* impossible. But the Comat power *is* an *inexhaustible outside source,* therefore the following fundamental proposition is true or false only in regard to the existence of such power. If the one proposition is accepted, it follows that we *have* to accept the other.

Free Energy?

Throughout the preceding chapters, it has been stated that given the validity of the theory, there is another extraordinary inherent implication which has to be addressed. This can now be approached by considering the analogy in Fig. 1.E. The diagram represents a conventional aircraft which incorporates some fundamental changes, and it is useful to remember that the thrust from a Syncomat unit is *constant,* regardless of the aircraft's forward speed. Also, in this step analogy, we will assume the aircraft isn't operating in a VTOL role.

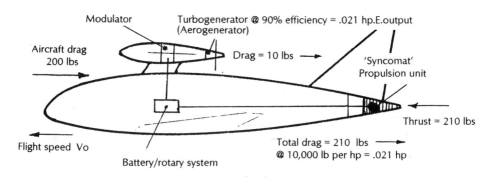

Fig 1.E. Schematic layout of Syncomat self perpetuating system fitted to an aircraft.

The EM Syncomat unit is first energised by an on-board storage battery, thereby providing thrust for the aircraft to accelerate and take off normally. On reaching cruising speed (Vo), the small, ram-air turbo generator is supplying electrical power to the Syncomat Modulator which, at ninety per cent efficiency, delivers the required pulsed energy to the Syncomat propulsion unit—say, 15.7 watts or .021 hp—which at 10,000 lb. thrust per hp (ascertained in the last chapter) produces 210 lbs. of thrust on the aircraft, which is exactly absorbed by 210 lbs. of aircraft drag *at that speed (Vo).*

Should for some reason—such as pitch changes—Vo decrease or increase a little, this loss—or gain—is expressed as electrical energy in the storage battery; thus, the system is virtually *self-perpetuating*. The only energy required is in starting the cycle, and even this may have been stored from the previous flight's excess energy. Note, this is only made possible due to the extremely low order of Syncomat required input power, and the fact that the main drive is utilising 'space' power.

Syncomat 'Free' Energy

In the foregoing transport application of Syncomat power, and elsewhere

Demonstration Syncomat generator built for a proposed BBC Tomorrow's World program which failed to be determined in mysterious circumstances.

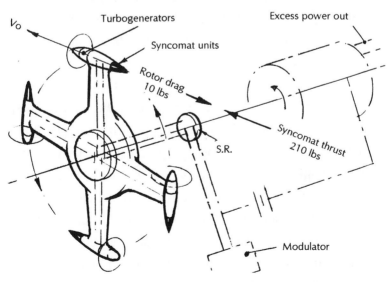

Fig 2.E. Analagous representation of a Syncomat 'free energy' power generator

in this book, I have offered the supportive wind driven sailboat analogy to help establish the somewhat difficult transition in changing direction from entrenched known physical laws—which demand much input for comparatively little gain—to a far simpler, high gain alternative. Now, interestingly enough, this simile is taking on a more prophetic nature, for we may now see there may be a far more profound aspect to it than we could have imagined.

We have seen how the arguments have led directly to what—in ordinary physical terms—appears to amount to perpetual motion, particularly where air transport is concerned. But if that is incredible enough, then the next logical progression can be awesome indeed. For if we can 'harness' the electromagnetic part of the Comat spectrum to move a machine at high speed through the air, then we are—in effect—generating power. Therefore, by implication, we should be able to similarly generate power on the earth in a ground-secured unit.

A more obvious, but not literal, method would be to develop the foregoing aircraft example further to accommodate, and we can do this with the aid of Fig. 2.E. In this, we delete the entire aircraft, leaving only a small nacelle, which houses the same turbo generator in the nose and the Syncomat unit in the tail.

This nacelle—or it could be one of several—is attached with its longitudinal axis tangential to the circumference of a wheel, which is mounted on a shaft running in bearings. In all other respects, the unit is identical to the aircraft version. Thus the assembly resembles an ordinary pelton turbine, only in this case the 'working fluid' is rather different in that wind, water or steam has been exchanged for latent 'space energy.'

The starting procedure is the same. First, the EM Syncomat units are energised via an external electrical source. When the tangential speed of the wheel has reached Vo, the turbo generators are then supplying the necessary electrical power for the modulator and the Syncomat propulsers, and the machine becomes self-sustaining. However, due to the drastically reduced drag and the relatively high thrust, the wheel will go on accelerating until the drag equals this thrust (210 lbs.). But coincident with this, the turbo generators will have supplied even more power to the propulsion units, so the catching up process will continue; unless the electrical supply is cut off, a runaway situation would develop, ending in overspeed and eventual centrifugal disintegration, as was discovered by Keely and others.

This is true only if the wheel shaft had been uncoupled. However, coupled to a specific load, it would reach operating speed Vo, and equilibrium, the excess torque being extracted as energy—*free* energy, which is *not* extracted from the earth's ecosystem.[2]

Fig. 3.E shows a simplified analogy of the same power plant, in which an inexhaustible supply of water represents the Comat energy. Once the validity of this simple process is acknowledged, the reader has but to accept the author's arguments for the existence of Comat energy and all that that implies and we arrive at the inescapable conclusion of the availability of an inexhaustible supply of clean and prodigious power for all our needs—including transportation. Of course, there are far more elegant translational processes which could be employed rather than a turbo generator system. Nevertheless, this illustration is sufficiently descriptive as a step introduction here.

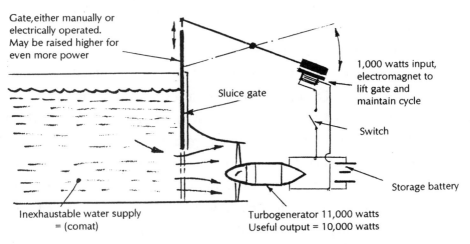

Fig 3.E. Simplified schematic of a water analogy of the Syncomat 'free energy' power generator.

Of paramount importance is the fact that if it requires only a few watts of energy to set the system going, it will be possible to supply every domestic home with its own 'free energy' conversion system. The units will be quite compact and comparatively inexpensive, requiring only a small amount of power to start, in fact much the same as small engine-powered generators are now. Thus, pylons and overhead transmission cables will be redundant, as will be power stations; for industry will use its own generation plants, thereby completely vindicating the age-old science fiction writer's domestic black box scenario, mentioned in the foreword to this book.

Finally, should some readers still remain unconvinced about the existence of Comat energy or the likelihood of our ever being able to use it, purely on the grounds that it smacks of something for nothing, may I remind them of the simple illustration—albeit not nearly so dramatic—we saw in Chapter 3, that of the simple, lighter than air balloon.

Extraordinarily enough, this classic simile involving winds was ironically appropriate, for shortly after penning these words in October 1987, our workshop was a casualty of the worst storm in the UK in living memory. The damage was severe, and some of the models and equipment suffered. Moreover, if the energy displayed by such air currents around the earth can be a potent enough reminder of the forces in nature, it may also prove useful to reflect for a moment on the mechanisms involved, correlated to the necessarily oversimplified theory set out in the foregoing chapters.

Contrary to the Laws of Nature?

In the foregoing context it is anticipated that there will be those who will apprehensively regard the employment of Comat energy as being "contrary to the laws of nature," and out of respect and sympathy for this view, I take this opportunity to reassure them with the following simile.

Given that the Comat exists, the situation today is exactly the same as in earlier times when mankind first became aware and took notice of the winds in the vast inexhaustible 'sea' of atmosphere he breathed. Next, he realised that by the process of providing sails, he could extract the latent energy in the winds to propel a boat. Then, by making the sails rotatable, he was able to extract the energy from the winds to power windmills and water pumps, etc. In any event, after extraction of the energy, the total volume of the earth's atmosphere was undiminished by this means, for even the immeasurably small energy converted into frictional heat, etc., was, in due course, ultimately returned to the vast overall ecosystem.

Today, in offering this book, I am merely stating my reasons for believing there exists not only a vast 'sea' of localised energy, but a 'sea' which permeates throughout all matter and the Cosmos. In this sea there are also waves of enormous inexhaustible power, waiting to be harnessed; and furthermore, I am also tentatively offering a methodology I have observed for harnessing this enormous power for all our needs. In a word, I am attempting to help design the 'sails' for our 'windmills' of the future, to consume relatively immeasurably small quantities of latent energy which will once again return to the vast natural resource totally loss free. As sure as the Almighty gave mankind the winds, so we have also been given the power of the Universe!

Intriguingly, it has recently been brought to my attention that, without over-stretching the imagination, there appears to be a degree of parity between the foregoing outlined principle and the symbolism in the footprint claimed by Adamski to have been made by a space visitor in the California desert sand in 1952 [Fig. 4.E (a) and (b)]. However, it should be understood that the author's analogous representations are offered with full acceptance of the fact that in this form, any resemblance to the ultimate development of the future is probably of similar magnitude to that of, say, Hero's steam turbine of 100 B.C.[3] compared to the multi-staged steam turbines of today.

We have seen that what we call gravity on the earth is relatively seemingly weak, yet it causes the atmospheric pressure, which in conjunction with differential solar heating, inertial and drag effects—caused by the earth's rotation, etc.—produces driving whirlwinds of herculean proportions that we call tornadoes and hurricanes. This is but one more visual measure of diluted Comat energy, the stuff which forms galaxies, suns and planets, which will always be there as long as there is a cosmos and that, given wisdom, is ours to use as our rightful heritage for all our survival purposes on this planet, as men have used the energy from the sun since time immemorial.

"Closing the Box"

One of the first observations on the introduction of such a new technology as Syncomat, is that—in keeping with the present general trend—the emphasis would be on electronic development, thereby replacing purely mechanical methods. For instance, apart from a plastic body, all of the 'functional' parts of a Syncomat car—including the power plant, stabilising system, controls, etc.—would be sophisticated electronics, with no moving parts at all. And this would be repeated in all branches of transportation and energy production.

To say the least, world-wide effects of Syncomat technology would be profound. There would not be one corner of our present civilisation left unaffected by it; *change would be total.* It would herald an end to inequality on a large scale. It would bring prosperity to underdeveloped parts of the world. Indeed, it would present no difficulty to fill this entire book (and many more)

Small version of an EM Syncomat generator described in the general summary.
(a)

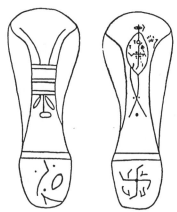

Fig 4.E. This sketch is a copy of footprints made in the California desert sands in 1952 by UFO contactee George Adamski's space visitor.
(b)

with the arguments *for* it, but as stated at the very beginning, there is much to be said *against* the acquisition of such technology *at this present time.*

A Social Plus?

Although it is true there will be countless other benefits and problems solved with the advent of such a new technology, there can be little doubt that transportation is by far the greatest contender, representing, as it does, a major environmental problem at all levels. Accepting that mankind's way of life cannot be totally changed overnight, motorways and highways will still be required. But while we have seen that road surfaces as they exist will be unnecessary for EM-type suspended transporters, it is obvious vehicular mass flow won't be reduced one jot. Indeed, for that reason, I suggest motorways may have to be replaced or widened by grassland strips.

As this will require even more land, land and house owners understandably won't be very pleased about that. However, bearing in mind that the chief complaint by those who are otherwise reconciled to having more motorways is one of noise levels—which will be absent with such a new technology—may help to redress the situation somewhat.

Not only will surface vehicles be lighter, cheaper and suited to underdeveloped parts of the world (thereby saving them enormous sums in road building costs), but this will be reflected by correspondingly lower levels of road maintenance normally required for *existing* roads all over the developed areas.

Not least, due to the fact that all such vehicles will be electrically powered (and stabilised in roll, pitch, yaw and heave in extremes of weather) they will be inherently compatible to receiving other tuned signals transmitted along motorways—as are currently street lamps—whereby vehicle speeds and guidance would be monitored and controlled, thereby bringing a halt to the horrific casualty figures at the present time. Moreover, use of fossil fuels for transport will be at a virtual end, ridding the world of one of the worst existing causes of extravagant energy consumption. And who knows, but fir trees in Germany might grow again!

An obvious advantage to Syncomat propulsion is, gone will be the days when we took it for granted that snow and freezing roads in winter months brought all surface traffic to a standstill. There will be no need for snow ploughs on snowbound roads and isolation of people living in remote areas, where even present helicopters sometimes cannot land due to 'white out' of the pilot's vision, caused by extreme snow ingestion into the rotors while in ground effect mode. With a Syncomat operated machine, such dire ground effects will be nonexistent, in that there will be *no* moving air streams (ejection mass) of any kind involved. Neither will old people continue to suffer and die in such conditions because they cannot afford to keep warm!

As I pen these lines, another more topical, but nonetheless important, environmental benefit of Syncomat power has become relevant, for the British government has just announced its intended U-turn regarding its policy of near-surface disposal of nuclear waste in favour of deep disposal pits.[4] With the pending election in mind, how much of this can be accredited to a vote-gathering manoeuvre, or the volatile reaction of the threatened communities, is largely a matter of conjecture. But what *is* certain is, once Syncomat power and transport is available, one of the first cleaning-up jobs that we shall be able to tackle will be the ferrying out of this disgraceful nuclear legacy by cost-effective space refuse disposal units. These will dump such waste of a bygone, unenlightened age into decaying solar orbits to be harmlessly vapourised, thereby ridding the world of a long-term, constant threat of radiation for our children's children yet to come! Indeed, such a cleaning-up job, with the aid of the solar system's vast natural incinerator plant, will no doubt be extended to include all kinds of otherwise unrecyclable rubbish. For the first time, we may have a solution for deviating the collision course of an errant asteroid by a cleaner process other than the nuclear bombardment alternative currently envisaged by some well-meaning, but not very practical, earth scientists!

A Social Minus?

If present day motorcycles, in the wrong hands, can represent a public hazard, then at least they are chiefly confined to public roads. But consider the Syncomat powered machine (described in Item 1, under General Observations later in this epilogue). Typically, this might have a weight far less than conventional vehicles, but we can assume that newly designed bodies would bring the gross weight to, say, around 570 lbs. In order to lift this weight to the hovering mode at the power loading quoted in the last chapter (10,000 lbs. per horsepower), the input would be a mere .058 hp. If, in addition, the machine was streamlined with a frontal area of some 6.5 square feet and a drag coefficient of, say, .08, an increase in power of only 5 per cent—absorbed in the form of forward vectored thrust—would produce a speed of over 100 mph—almost silently. While an increase of 10% of this minuscule power to about .062 hp would produce a speed of an *impressive* well over 130 miles per hour! Over hedges, trees, houses, lakes, *anywhere*. How do you *realistically* control a problem like that?

Again, if you were the owner of a 'super car' of the kind presently described, it will have a format similar to that shown in the Prologue. It will be equipped with small retractable tandem bogey wheels for ground handling and two retractable lateral support pads. With power on, but not lifting all the weight, the lateral pads will be retracted and the vehicle will remain vertical due to the automatic stabilising system, thus in this mode it will be easily manoeuverable. But in operational mode, the super car will be raised above the surface (water or land) to a few feet. However, the power required to hover it at a few feet will be identical to the power required to hover it at 10, 50 or 100 feet or more. Therefore, by implication, such a machine has inherent VTOL capability much the same as a helicopter. But it could do the same in the vacuum of space! Therefore, as all aircraft (or hovercraft) are by definition 'air supported vehicles' (ASV's), Syncomat propelled vehicles are neither. Legislation should be interesting! Some kind of legislation would have to be introduced, but how?

I wish I didn't have to say this, but the fact is, the standard of driving on our roads today is nothing short of appalling. The *real* miracle is, that there aren't even more accidents than those shown in the Appendix. Most people—not just the odd few which we might normally expect—haven't the slightest idea how dangerously they drive, i.e., *far too close at far too great a speed.* The police emphasise this fact *every week.* As I pen these words, there have been two other multiple vehicle pileups resulting in people being maimed and killed, due to driving at *70 mph* on a motorway in *ice and fog!* OK, so people *are* irresponsible, but the mind boggles at the prospect of what things could be like if they all had access to a super car! Clearly some kind of control or widening of motorways with grassland belts *is* indicated. Travel should—and could—be a pleasure, not a nightmare!

Yet another instance: at the present time, almost anyone who can afford a helicopter can obtain a licence to fly it within a matter of a few weeks. Apart from the usual restrictions (minimum height over towns, and air traffic approach to airports, etc.), there is—as yet—no limit set on 'ceiling,' which of course is automatically governed by the fact that air supported vehicles—including helicopters—require an atmosphere in which to function; for that matter as do conventional air breathing engines. But what if neither of these limitations existed?

By way of illustration, let us imagine Syncomat units have been built, tested and certificated. A conventional, specially pressurised small passenger aircraft is fitted with a series of these, and can take off and land vertically. Put into hovering mode, it behaves exactly the same as a helicopter; thus, the above newly fledged helicopter pilot could fly the machine vertically to any height, beyond the atmosphere and into space, and over *any country!?*

What Limits?

More specifically, let us take a four-seat family Syncomat powered car. The driver and three passengers might weigh around 680 lbs. As we have seen, the vehicle would be light—around 750 lbs.—so the gross weight would be 1,430 lbs. Therefore, in order to climb vertically, the lift would have to be something more than this, but say 2,860 lbs., which at 10,000 lbs. thrust per hp gives required horsepower as .286, or just 213 watts! The *average cyclist* can generate more than this at around .35 hp; moreover, any machine that can lift twice its own weight has a 2g acceleration capability, which is very impressive indeed! Therefore, this family car, if necessary, could be started up manually, lifted off the surface to hover at, say, 5 to 100 feet, and if so kitted, accelerated to the aerogenerator's self-sustaining cruise speed.

The same machine, equipped with wings, could take off vertically, rotate to an angle of attack — jump jet style — and accelerate at that inclination at a *very* useful speed. Not having an air breathing engine, it could travel on upwards exactly as the little craft described in the Prologue.

By all our *established* physical laws, this may sound very farfetched, but remember, we are not extracting energy *from* matter to supply other kinds of limited energy *to* matter. We are using energy from matter to extract a vast amount of energy from *space*. If that *is* acceptable, what about this. The same vehicle, instead of returning to Earth, could carry on towards the Moon! In airless space, the turbo generator would be inoperable so it would be operating on the on-board self-generating unit or, if necessary, the storage battery. Even a standard car battery gives around 40 amp hours, which at 12 volts equals about 480 watts per hour. Two batteries would give about 960 watts which, at the above 213 watts requirement, gives 4 to 5 hours duration.

Now, the distance to the moon is approximately 238,857 miles, giving a mean distance of 119,428 miles. Instead of being hauled to the moon in one short blast and then coasting on—rocket fashion—the above vehicle could go on accelerating comfortably at 1g (32.2 ft. per sec. per sec.) to the halfway mark—119,428 miles—then decelerate at this same (1g) rate, arriving at the moon after only 3-1/2 hours. The total return trip would take 7 hours, during which time the occupants of the vehicle would not be exposed to zero g (weightlessness) conditions. We saw above that, with two batteries, the duration would have been only 4.5 hours. Thus, for some of the return trip, the pilot would have the choice: battery power, on-board self-generation unit, or, as a friend of the author so lightheartedly put it many years ago, "he could pedal himself down." The whole point of the exercise being: to illustrate a measure of the comparatively minute powers involved and the fact that difficult as it might be to believe, the time could arrive when space flight will be common! However, should there still be doubts remaining about this, then consider the following: for centuries, travel across the seas of the earth had been left to a comparative few. Today, millions of people cross those same oceans yearly, by air or by sea; apart from agreed international limits, nobody says they can't.

Similarly, at one time not so long ago, motorcars were considered to be only a whim for the mad or the very rich; now the world is infested with them. Once a cheap and safe method of space travel has been achieved, who is going to tell us that we cannot use the oceans of space around our planet? By what government of what country, and by what measure of legislation, can this be *realistically* sorted out? When—as previously stated—at the present, if one has a pilot's license, you can fly an aeroplane as high as you are fool enough to try? But suppose it was no longer a risky thing to do; and let there be no doubt about it, that time could come! The reader is urged to remember that, within this context, we are in exactly the same position of disbelief as were the sceptics about the future of the motorcar.

Certainly, the above moon trip may, at present, sound rather unlikely, just as did predictions of travelling in 'horseless' carriages at over 20 miles per hour only a generation ago. We should not forget the exponential rate of development discussed in Chapter 2. Also, one thing is very clear: I have never met one competent aerospace engineer who *seriously* believes that space flight by rocket is ultimately going to be the *only* way!

Political Issues

Lest the reader has any doubts as to the world's capacity to cope with Syncomat energy and transportation, the following typical example may serve as a guide.

Today, nobody would try to prevent the sale of a solar heating engine, for the parabolic mirror is already a long established device, so is the heat exchanger and every other part. Therefore, we already have an accepted 'free energy' source to a very limited degree.

Just like the parabolic mirror, a Comat converter could become a well-established device, so might the rest of the energy developing unit. The one situation is identical to the other. Yet there can be little doubt that, given the overnight availability of the latter, one can without difficulty imagine some of the frantic and lunatic objections which would be dreamed up by bureaucrats in order to prevent it! There would be all sorts of excuses and reasons produced which, at the end of the day—in real terms—would amount to, "the world's economic system is too deeply entrenched for such a drastic change." This elementary example is but one of many which could quite easily fill this entire book.

Pollution

As already discussed, and more statistically shown in the Appendix, the greatest producer of atmospheric pollution is the motorcar. With the introduction of Syncomat power, all such fuel burning internal combustion engines—including lawn mowers, motorcycles, etc.—would eventually become illegal, as would noise from aircraft, including helicopters, etc., to raise but a few instances.

Industrial

Another reason for including statistical data is to impress the reader with the enormous degree of capital investment and energy that is consumed by industry. In fact, we need look no further than the manufacture of motor vehicles to be convinced of that. Practically every car owner is aware of the future-oriented programmes, which involve thousands of millions of pounds. In the present context, the question has to be asked, what is going to happen when the designers and investors finally realise that the era of the motorcar—as we know it—may really be over? And again, while many of the work staff at any of these establishments would be delighted to have an X15 Falcon, would they be willing to lose their jobs because of it? Currently, over fifty per cent of all natural and synthetic rubber is used in the manufacture of car and truck tyres, and that industry alone is one of the major employers of people all over the world. It's all very well to assume that they would be allocated different jobs in new kinds of industries, but there would be many other trades and industries involved, and the change couldn't be made overnight. So governments would have to be more farsighted and get their act together and help people to prepare for future change.

We all want progress but not, it seems, if it means change; the last miners' strike in the U.K. is evidence of that. Naturally, this is understandable. Why should people have to leave their homes and change their occupation? In the normal run of events, of course, they shouldn't; however—no doubt difficult for some to believe—the future scenario we are talking about is far from normal. Indeed, as I said in the beginning of this book, the situation may eventually become of far greater magnitude than a third world war. In which case, with all due respect, this author is tempered by memories of the time during the Second World War when, as a youngster, having spent a year digging up delayed action bombs, I was discharged from the army before being compulsorily requisitioned by the Ministry of Labour to take up employment hundreds of miles away from my home and family, whom I could therefore only visit for the odd weekend.

The whole point being, I accepted change then because the country was at war and the future scenario I am reluctantly portraying is no less than that! If this proves to be so, ultimately our very survival may depend on change, of that there can be little doubt. We may *have* to accept it and hope that people will be equal to it and control it, for the betterment of all concerned. Also, governments will have to diligently control those who would stop survival-oriented change due solely to vested interests. Not so long ago, governments also may have considered they had good reasons to prevent such development, as would the large oil industries, etc., but self-evidently the time is not so far off when there isn't going to be sufficient oil left to be monopolised by anybody. *We have to begin looking for something new.*

Trade Unions

The author has spent most of his life in industry and been a trade union member for much of that time. I have been grateful for the times when, through trade union action only, I have been granted a well-earned rise in salary I would not otherwise have received. Therefore I feel qualified, within the present context, to make an unbiased judgment here. That is: if the trade unions were originally formed as a result of the unquestionable tyrannical nature of wealthy, powerful employers of yesteryear, then sadly I have witnessed the swing of the pendulum, in that the single-mindedness and cussedness of some trade unionists has become a close contender.

For instance, in a large establishment I have seen the success of a very important project seriously threatened due to the fact that strike action was instigated. A works cleaner showed he could build for about £150—from scrap—a tank towing rig which was currently costed at some £5,000! That project was extremely important to us *all*. This is power taken to stupidly irresponsible extremes!

Faced with propositions concerning our ultimate future on the one hand, or losing a job—albeit in exchange for a different and better one—on the other, we know from experience what some trade union officials would say. But reassuringly, the author remembers those World War II years and I know that when faced with a *real* threat, the *ordinary* people—freed from well-meaning but quite misguided officialdom—rally together and *do* the *impossible*. There is little doubt, sooner or later they are going to have to do it again.

In terms of the exponential rates of development we saw in Chapter 2, there are grounds to suspect that the natural rise—and sometimes eventual decline—of a civilisation is a kind of race, in which a species grows and develops, obliviously gobbling up the planetary resources until they are depleted. If, during that time, the technological/cultural development has kept a balanced pace, then the species has the technology and the right to extend its existence and further development into deep space. However, should that balance not have kept pace, leaving a negative deficit, then the species will neither have the technology nor the right to interplanetary expansion, and, due to some reason or another, it will decline and cease by exactly the same exponential rate. For what it's worth, I feel there is evidence to suggest that this has happened before. It will be a great calamity if it now has to happen again.

It should be made clear that the concept presented here cannot be interpreted as an immediate panacea of all our energy and transport problems. All we have to do now, is drop everything else and research Syncomat power. As stated at the beginning of this book, investigation should begin now, alongside the other alternatives. If, in so doing, it *is* successfully developed ahead of all other efforts, then that will be one large bonus and we shall have our panacea.

Again, I would ask indulgence of those readers who instinctively baulk at some of the issues stated herein; as an unbiased engineer, I admit I have, also. Indeed, it would be strange if we didn't. But it has always helped me to remember those who, in decades past, also baulked at 'preposterous' propositions made by a few, the results of whose dedication you and I are now apt to take for granted.

Comat energy is nothing new, in the sense that it has always been here. Its discovery and utilisation will represent an extension of known physical laws, rather than prove at variance with them. We have only to identify and tap it, as our forebears discovered and tapped oil before us. It is hoped that if (or when) we do, we shall have the good sense to use it wisely for the benefit—rather than the desecration—of our planet and all mankind.

These and many, many other issues—such as the passage of time, rumours from around the world of other claims for energy 'breakthroughs,' global warming, the energy crisis, political and attendant problems, etc., etc.,—have all contributed in prompting the author to publish these findings, albeit somewhat prematurely. They represent a true account of my participation in some known, and some hitherto unknown, research. In addition, the reader is advised not to be misled by some people of orthodox academic rank who (if ever my findings are verified and put to good use) will try to adopt the "of course we knew it all along" attitude. Having spent so much of my life—and suffered some of the attendant consequences along the way—I can state unequivocally that *currently* far from *knowing it,* they would vehemently deny its truth. They belong to the 'bumblebee can't fly' brigade and should be cautiously avoided.

As stated at the beginning and reiterated since, we are not talking about some distant future or pseudo science fiction, we are talking about now and fact. Whether in the East or West, someone, somewhere, sometime, is going to develop such technology as that described in this book. Whether there are those who will believe it or not, time will vindicate much of which has been said here. I am merely putting it on record for those who would share its significance. Although I have great confidence in the authenticity of the foregoing basic hypothesis, at this time it would be premature of me to presume as much for my own interpretation of it, save to repeat, I suggest it will be discovered it is close to agreement.

To cosmologists and particle physicists, I extend my apologies for, somewhat amateurishly, venturing into their chosen domain with a simple visual model. Although lacking in further elaboration and mathematical assurance I suggest may prove useful, I did so that I may have succeeded in contributing a little towards a greater understanding of the intimate relationship between space, matter and energy and, who knows, helping a little towards sorting out some of the man-made chaos.

Self-evidently and sadly, there is a proportion of our society who are incapable of carrying such a prize as that predicted herein, which may not belong to their order or time. But what of the alternative? *Is* the conflict prophesied in the Box of Pandora nearly over, or is it just beginning? We may soon find out.

General Observations and Some Fundamental Applications

There can be little doubt that two of the most significant problems besetting the modern world are mobility and energy. So although there are many applications for a new technology of the type discussed in this book, whether intentional or not, these are the ones which attract most attention. Accordingly, the following is a logical anticipation of only a few of the issues which will be raised.

1. Most domestic cars will not require conventional road wheels or suspension, neither will they have to rely on a conventional braking system, due to the fact that Syncomat drive units can be instantly reversed, thereby providing controlled, efficient, skidless braking. All things considered, cars will be about 50% lighter than they are now [Fig. 5.E (a)].

2. Every domestic car will have VTOL capability and, in some instances, be considered potential spacecraft. They will also operate as submarine vehicles, silently [Fig. 5.E (b)].

3. Small, two-seat hovering vehicles, similar to present motorcycles, will be available [Fig. 5.E (c)].

4. Heavy freight vehicles of similar size to—though more slender than—present large hovercraft, will be employed.

5. Main roads will still be used for close surface operation, but with extended grassland belts on either side, greatly reducing traffic density on these roads. Much of the remaining traffic will be at controlled higher altitudes [Fig. 6.E (a)].

6. All railways and airport runways will become obsolete.

7. Generally, air pollution due to all vehicles will cease.

8. Such transport will be a boon to the underdeveloped parts of the world. For instance, trackless vehicles of army tank mass and proportion, operating above ground, could easily penetrate dense forests to attend fires, rather than kill and maim [Fig. 6.E (b)].

The Lunar Surface vehicle shown in Arthur C. Clarke's *2001: A Space Odyssey* was extraordinarily prophetic. Due to the relatively enormous power loading of Syncomat impulse units, the question of specific application becomes redundant, for with such available power, *all surface* vehicles would logically operate above the interface, be that land or water. Conventional road wheels and suspension being obsolete, only small in-built servicing castors would be necessary. Thus, with this kind of propulsion system, we will have arrived at just

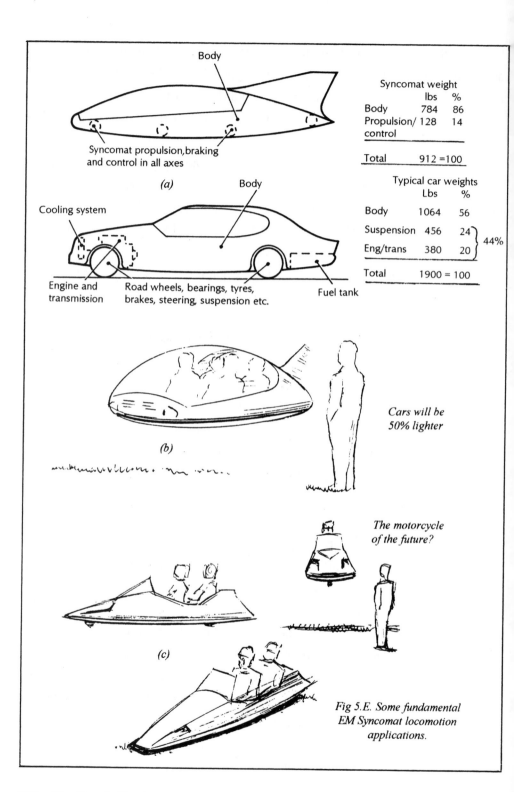

Syncomat weight

	lbs	%
Body	784	86
Propulsion/ control	128	14
Total	912	=100

Typical car weights

	Lbs	%
Body	1064	56
Suspension	456	24 } 44%
Eng/trans	380	20
Total	1900	= 100

Body

Syncomat propulsion, braking and control in all axes

(a)

Cooling system

Body

Engine and transmission

Road wheels, bearings, tyres, brakes, steering, suspension etc.

Fuel tank

(b)

Cars will be 50% lighter

The motorcycle of the future?

(c)

Fig 5.E. Some fundamental EM Syncomat locomotion applications.

Large grassland belts on either side of existing motorways will help to disperse controlled one way traffic. Heavy freight vehicles being restricted to existing centre road. (a)

Fig 6.E. Massive Syncomat trackless vehicle combating forest fire (b)

one kind of transport vehicle. No doubt freight transporters will still look like transporters, as will the cars of the future retain individual styling, even so our long need for diversity in this area will be over.

In-built Stability

Syncomat drive units will be so conveniently small that it is logical to divide the total installed power into a number of completely independent units, while the total power can be doubled or even quadrupled, to anticipate adequate 'redundancy' measures [Fig. 7.E (a) and (b)]. Some readers may be unfamiliar with this aerospace invented term, but the argument behind it is logical enough. Broadly, it can be analogously put that if, say, 10 identical motorcars powered by

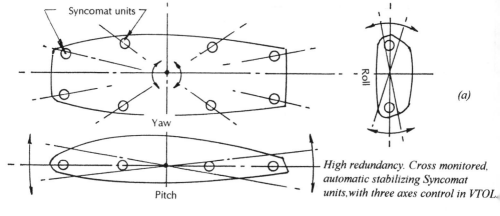

Syncomat units

Yaw

Pitch

Roll

(a)

High redundancy. Cross monitored, automatic stabilizing Syncomat units,with three axes control in VTOL.

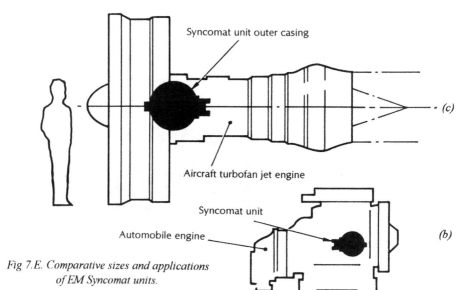

Syncomat unit outer casing

(c)

Aircraft turbofan jet engine

Syncomat unit

Automobile engine

(b)

Fig 7.E. Comparative sizes and applications of EM Syncomat units.

the same make of engine, with an adequate fuel supply are being driven along a motorway at varying speeds, then the chance of all ten engines stopping simultaneously—for some reason or other—is astronomically unlikely. In a word, there is safety in numbers.

Because of the electrical nature of the proposed system, it is capable of extremely rapid orders of thrust (unbalance) changes—as are all common electric motors. Therefore, several units working in unison, with computerised solid state rate gyro systems, will quite easily and effectively accommodate dynamic and static stability, in heave as well as the three main axes. Years ago I visualised such techniques, it seemed remote indeed! Today, all the necessary hardware is available, thus a 'lock-on' process can be installed in such vehicles to render them completely immobilised whilst in the hovering mode, as well as providing a unique braking and control system.

Aerospace

Insofar as the emphasis will be on above surface travel at all altitudes, the earlier quoted thrust figures produce some interesting additional specifics, particularly where aviation is concerned, for unlike air breathing and expelling devices—such as propellers, rotors and pure jet propulsion engines—the thrust power of a Syncomat unit is *constant, regardless* of forward speed. Also we note that it is no respecter of 'up' or 'down' orientation, due to the fact that Comat rays are omnidirectional.

9.   All aerospace vehicles will have VTOL capability and will be silent in takeoff and landing modes.

10.   There will be many smaller airports distributed around the country, reducing transit times to an acceptable level. Existing large airports will be re-sited and/or reduced in size.

11.   With auxiliary stored electrical power and the aforementioned, high redundancy factor, all aircraft will have a very safe power reserve.

12.   As Syncomat propulsion units will be completely silent, within the aircraft there will be only slip stream noise.

13.   There will be no moving parts, but for the exception of an aerogenerator for rotary storage and cooling systems (if required).

14.   There will be no vibration, and no risk of bird ingestion into intakes.

15.   Aircraft will still be required to be aerodynamic, but more in the sense of reducing drag. Lift induced drag may eventually become zero, thus, aircraft as we know them will cease to be. This includes helicopters and airships.

16.   Being far smaller than any existing power unit of similar power, Syncomat impulse units will fit conveniently into any conventional aircraft [Fig. 7.E (c)].

17.   Syncomat space vehicles will be able to go into orbit, or if more convenient, ascend to the required altitude and hover for many hours or days if necessary, the propulsion units being powered by the auxiliary means. Note, in the latter mode, zero gravity and space sickness will become problems of the past.

18. Descent from such a trip will be a controlled 'flydown' situation, thus there will be a complete absence of aero-kinetic heating—as we saw earlier in the Prologue story.

19. The enormous volumes of expensive and volatile fuel carried by present aircraft will cease to represent a major flight hazard. What fuel is carried will be minimal, optional and certainly disposable in the unlikely event of an emergency vertical landing.

20. Among other things, the theory also predicts that on the domestic scene people will virtually be able to own private VTOL flying *homes*. They will have a circular planform and a lenticular profile (which offers the maximum accommodation space), and be so stable that the occupants will be able to walk around or have a comfortable meal in flight, in much the same style of luxury enjoyed by passengers of early large airships. These vehicles will also be silent and without fins or control surfaces. People will be able to travel anywhere—at around 400 mph or more—to other parts of the world for extended holidays, alighting on water or land, etc., etc.

And finally:

21. There is evidence to show that such technology as Syncomat *can* happen.

22. There is evidence to show that something of the kind *has* to happen.

23. There is evidence to show that the present society may be incapable of handling the attendant change.

24. There is evidence to show that there may not be a meaningful long-term alternative.

Self-evidently, whatever technological changes may occur, there are sound reasons why they should not be too revolutionary; we have a need for *gradual* changes, rather than risky 'overnight' ones. Considering the foreseeable dire need, it is to be hoped that rather than deterring investigation into this exciting area, meaningful encouragement is offered in order to ensure an acceptable, smoothly controlled transition when it *is* necessary. Indeed, for all we know, somewhere that may already be the case.

Due to the significance of Item 23, together with other considerations, regretfully, I have not continued meaningful investigation in the Syncomat project for some time. I have been fully reconciled to another, no less exciting but mechanistic interpretation for transport locomotion, which inherits many advantages of the foregoing.[5]

Several years ago, Timothy Good asked me if I would receive a phone call from a brilliant young Israeli physicist who was keen to know about my work. While in the process of penning this Epilogue and thinking of an appropriate ending, I heard from this young man who, regretfully, was rather appalled to learn of my decision. In no uncertain terms, he said he thought my attitude was rather irresponsible, for didn't we desperately need something like Syncomat to save the planet? etc., etc. I had heard it all so many times before. Indeed, as a young man of his age, it was exactly the stand I used to take. As politely as I could, I urged the young man to consider very carefully the ramifications involved. Then several weeks later he phoned me again, as I thought, to reinforce his point of view. But I was quite wrong. He said that not only had he done as I suggested, but he was phoning me to say he had now completely changed his mind, that there is after all a time and place for everything, and for Syncomat that time wasn't now.

I have included this short narrative because I know there will be many other young people who, having read this book, will likewise need time to think. I urge them to remember, I was the young man who, years ago, used to address my audience with much fervour and urgent persuasion for us to "tap the energies of the universe" overnight! Nevertheless, it is my sincere hope that someday, somehow, mankind will mature sufficiently to recognise and utilise this wonderful resource we have inherited ... *as I have no doubt others out there are using it today.*

1 Illustrated in Appendix 2.

2 In this analogy there would, of course, be mechanical wear; therefore, strictly speaking, it should not be interpreted as a 'perpetual motion' machine.

3 Hero of Alexandria.

4 'Nirex.'

5 Currently being investigated by an international consortium.

APPENDIX 1

SUPPRESSED INVENTIONS
of the Corroborative Kind ?

It has been considered appropriate to bring to the reader's attention some otherwise little-heard-of facts relating to 'borderland' energy research projects of the past, together with some of the bizarre associations surrounding them. Also, from what has already been discussed here concerning a cosmic energy source, it will be apparent that if the concept *is* valid, we shouldn't be surprised if other claimants included reference to an antigravity phenomenon, for as we have seen, according to the foregoing reasoning, the two factors would be manifestations of a common cause. Of the known cases, it is the work of John Keely in which such phenomenon was recorded and, due to the fact that it took place so long ago, it has been dealt with here at some length as a small token of acknowledgement to this brilliant and lonely man.

Although in this and several other instances it might appear that some kind of preventative action may have been involved, some might argue that there could also have been a more benign socially orientated motivation. Be that as it may, in no circumstance can such extreme measures ever be justified—nor go unrewarded as was revealed in Chapter 6.

The Keely Motor Enigma

About 1866, John Worrel Keely of Philadelphia, USA, claimed he had discovered a new force in nature which he succeeded in harnessing. The United States government was interested because, in a small machine, Keely produced sufficient power to fire a small cannon, but the inventor refused to reveal his secret to anyone. Scientists—including Professor Leidy, Dr. Wilcock and others—were confounded and businessmen panicked. None understood the mysterious power of this strange machine and nobody had offered an acceptable explanation of how John Keely created power by 'disintegrating air.' Among those who pleaded with him for the secret were some of the country's leading scientists, engineers and inventors, including Thomas A. Edison. To this day, physicists and engineers have been unable to explain how he generated the inexplicable force. Many agreed that Keely actually produced energy.

In answer to a question put to him by a Major Seaver, Keely said:

The idea came into my mind from where I cannot tell. Perhaps it first came from a craze I had to study the magnet to attempt to solve what mysterious power was that enabled it to attract steel and iron to itself.

Where does this really tremendous amount of energy come from? By what inscrutable process does the mere magnetization of a bar of iron make of it a machine for the transformation of energy, even more, a perpetual creator of force?

It came into my mind that there was a hidden process going on of some kind, energy going into a magnet and flowing out of it all the time doing work—energy of some form.

Where did it come from, gravity, atmosphere, solar rays, earth currents? Who can say?

The mere fact of the magnet carrying its load proves conclusively the constant flow or positive action of a sympathetic force, the velocity exceeding millions of vibrations per second.

The basis of Keely's work was vibration, and it should be noted his sceptics loudly declared this was nothing but a front for his fraudulent claims. In performing one experiment, employing vibrations of an extremely high order, Keely wrote:

The highest range of vibration I ever induced was in the one experiment that I made in liberating ozone by molecular percussion, which *induced luminosity*,[1] and registered a percussive molecular force of 110,000 lb. per square inch, as registered on a lever constructed for the purpose. The vibrations induced by this experiment reached over 700,000,000 per second, unshipping the apparatus, thus making it insecure for a repetition of the experiment. The decarbonised steel compressors of said apparatus *moved as if composed of putty*.

Keely's theoretical explanation was that *"the force that controls planetary suspension is a sympathetic relationship of 'ether'* and that he had tapped the current or *polar stream*, which is *negative* and *positive* in its attributes and may be reached by and made to work for mankind by *sympathetic vibration*."

The public story of the Keely *engine*—which was to rock the country—began on a dreary autumn day in 1872, when the inventor met secretly with a group of bankers, physicists and engineers. For six years before this, Keely had experimented with vibration as a means to produce power. When these experiments exhausted his capital, he went to Charles Collier, a leading patent attorney. After Collier saw several demonstrations of the machine, he suggested inviting bankers to get the necessary financial support for further and greater experiments. Keely opened the meeting by coming straight to the point. "In considering the operation of my engine you must discard all thought of engines that are operated upon the principle of pressure and exhaustion, by the expansion of steam or gas. My system is based and founded on 'sympathetic vibration.'" The scientists Keely had brought were men of high standing and unimpeachable integrity, and they corroborated his statements. They had seen him develop 20,000 to 32,000 pounds pressure per square inch, in the first experimental stages of his 'ether smashing small machine.'

Before the meeting broke up, a committee was appointed to go to Philadelphia to study the inventor's achievements. Simultaneously, he was given a token cheque by the bankers for $10,000 to continue with his experiments until the committee made its report. Keely agreed to demonstrate everything he had achieved to the committee's satisfaction, *provided they did not ask to see inside his machine.* With the $10,000 Keely bought new machinery and continued with his experiments. Periodically the financiers, engineers and physicists went to Philadelphia to watch his progress.

Psychokinetic Triggering?

At one time the shareholders of the "Keely Motor Company" put a man in his workshop for the express purpose of "discovering his secret." After six months of close watching, one day the man said to Keely, "I know how it is done now." They had been setting the machine up together and Keely had been manipulating the switch which turned the force on and off. "Try it then," was the answer. The man turned the switch and nothing happened.

"Let me see you do it again," the man said to Keely. The latter complied, and the machinery operated at once. Again the man tried, but without success. Then, Keely *put his hand on his shoulder* and told him to try once more. He did so, with the result of an *instantaneous production of power.*

Then Keely announced that he had succeeded in overcoming the force of gravity and demonstrated evidence of this claim before a special committee. The incident was described in the *Philadelphia Evening Telegraph* on April 13, 1890. The report stated that Keely used a model of an 'airship' which weighed about eight pounds, to which a 'differentiated' wire of silver and platinum was attached, "communicating with a *sympathetic transmitter."* It is claimed that the model rose, descended or remained stationary midway, the motion being as *"gentle as that of thistle-down floating in the air."* After this meeting the bankers suggested that a Keely Motor Co. be organised so they could receive stock. Keely got one-seventh of the 20,000 shares issued, but no effort was made to sell the stock publicly.

Keely said he had 'hitched' his motor to the *polar stream* and that his *push engine* could be bolted to the front of a streetcar and once the *chord of the mass* was struck and the *correct frequency* was obtained, the streetcar would accelerate "as fast as you please." He claimed that no gearing or wire connections of any kind would be necessary between the push engine and the wheels of the car. He also claimed that the disintegration of three drops of water by means of his new force, furnished power enough to push the car along for no end of time.

An eyewitness of the experiments in the Keely laboratory said that by this force, a steel shell weighing several tons was raised from the floor to the top of a tripod six feet above the floor, there being no mechanical apparatus in view but a small box, of the 'push engine' variety, connected by a wire with the steel shell. The push engine was described as being cylindrical in shape and eighteen by ten inches in diameter.

At this point Keely Motor Co. stock was offered to the public by the corporation which the bankers controlled. Within a short time the stock jumped 600 per cent. Frenzied buyers paid $2,000,000 for shares. Speculators cornered the stock and the value skyrocketed even more. Then at the height of this world-wide fever, in 1898 John Keely died.

But the world still wanted to know the secret of the weird machine and a final committee of nationally known scientists and engineers was appointed to investigate it. They dismantled the machine and for the first time saw the inside.

A report said, "There was a maze of tubes *crossing and crisscrossing* each other in *every direction* and connected to what seemed to be everything else." They traced this maze of pipes and reported back that there was absolutely nothing they could see which would indicate the ability of the machine to produce the enormous power that Keely had repeatedly demonstrated. They asked, "Where did the power come from?" It came from the machine, all the scientists and engineers agreed on that, they had seen it come from the machine. But how? The only man who could give the answer was now dead and the secret of his extraordinary power was buried in the grave with him.

There can be little doubt that here is an instance where such a discovery would have been centuries before its time, for its successful application would have required modern supportive technology. One example will suffice. Imagine the world having an anti-gravity technique *before* the Wright brothers first flew. They would have known nothing about wind tunnels and aerodynamics and even less about space flight. Clearly, there *has* to be a natural sequence in technological development, as I stated in Chapter 2.

Even a very brief perusal of Fig. 1.A1.—despite the poor reproduction—reveals to any engineer the wealth of intricacy which had gone into the manufacture of Keely's machines and apparatus. Enough of it can be identified and, whether sceptics like it or not, what we see is entirely consistent with sonic and vibrational work. In other words, evidence of experiments in the domain that Keely said he was experimenting in; no more, no less. Therefore, this much we know: he was sincere in that part of his claims, and he was hardly likely to have constructed complicated devices over a period of *twenty years* merely to fool people. At least he knew what he was trying to do.

His description of the beautiful colour effects given off by some of the experiments was a predictive description of sonoluminescence[2] by no less than one hundred years! This is hardly likely to be a case of coincidence-favoured fraud. Apart from the testimony of scientists of the day, as we have seen, it is logical to believe that any claim to obtain energy from space inherently implies a means of so-called anti-gravity. Though note, Keely presented the latter as something of a secondary issue; this also lends forceful argument in his favour. His descriptive terminology—which he had to invent—regarding his theory of

Keely's Globe Motor and
Provisional Engine..

Sympathetic Negative
Transmitter.

Keely operating the Liberator
to drive the Musical Sphere.

*Fig.1.A1. Keely backers assumed his motors would 'Power the World'. Modern technology
might well benefit from an investigation into Keely's claims as today's science becomes
more aware of hidden potential in nature.*

matter and space, is close (too close) to the Cosmic Matrix theory to be purely coincidental. Clearly, rather than a charlatan, this lonely man was centuries before his time. In an age without electronics and computers, he had burrowed away to find a purely—by today's standards—antiquated mechanistic, but beautiful, solution to energy and gravitational problems.

The reader's attention is especially drawn to a report by Cyrus Field Willard, in which he described that, together with a Dr. Franz Hartmann, he visited Keely in 1896. He said:

> Mr. Keely's laboratory contained a number of instruments which were unfamiliar to Dr. Hartmann and to myself and of whose use we were ignorant. In going through one room in which there was a bath-tub I noticed an oblong block of granite about 12 by 14 inches and 2 inches thick. Alongside it was a small mound of almost impalpable powder. I picked some up in my hand and noticed small bits of biotite such as is usually present in granite. I asked Mr. Keely, 'How did you get this powder so fine?' 'Oh, by just withdrawing the power of cohesion,' was his astonishing reply.
>
> Mr. Keely then showed us some things that looked like empty shotgun shells, which he said he used in the globe arranged along certain lines of force to receive certain vibrational impulses. In one of these we saw a peculiar glistening sand, which, he told us, received the impulses. He poured some into my hand. It looked different from any silica sand I had ever seen and was more like carnotite or scheelite with a crystalline glistening yellowish brown colour, which I examined very closely while Mr. Keely smiled as if to say 'I defy you to find out what it is!'

Before closing this account of John Keely, I have given way to a tempting little incident which is really of no import, yet in some way it may help in drawing some readers a little closer to accepting the validity of his claims.

In 1960, I was invited by a brilliant young computer scientist, Dr. John Wright, to give a talk to students at Southampton University. Knowing how visual analogies can help keep an audience awake, I had built a little rig to simulate the idea of gravitational control, as shown in Fig. 2.A1.

This consisted of a powerful permanent magnet—which in the analogy represented the Earth—'attracting' toward it a little circular 'spaceship' formed by a simple coiled heating element. I wanted to show the difference between a rocket fighting this 'pull' by the ejection of a mass spewed out of its internals

When there is no current flowing in the iron coil, the permanent magnet attracts it. This is analagous to the state of gravity.

Pivoted balance arm

Iron coil

Permanent magnet

Heated coil

When current is applied to the coil it heats up (molecular vibration) and the magnet loses hold over it. Being analagous to 'gravitational transparency'.

Fig 2.A1. Author's permanent magnet and heated coil analogy illustrates a kind of selective gravity.

(Newtonian reaction), as against the loss of attraction in the coil by the simple process of heating it by passing a current through it *from within the system*. As is well known, the magnetic field is interfered with by the red heat and, in the analogy, the magnet's 'pull' (gravity) is therefore annulled.

Now there is nothing new or very profound about this, but while demonstrating the analogy that day, I suddenly saw how it was so very near to the claims of Keely, for he believed that gravity could be overcome by certain orders of *pure* vibration, and heat is exactly that, *vibration*! I wasn't talking about Keely that day, but all at once, he seemed to be very near. In my mind I remembered that other day, in the advanced projects office at Napier's in 1956, when I had put my searching question to Dr. Morley, followed by his frank answer, "Often as not we can't see the wood for the trees."

Contrary to popular belief at that time Keely was never really exposed as a fraud. Does the whole affair *really* remain a mystery?

The following footnote to this section on John Keely should be read in conjunction with the section in Chapter 9, headed "Format for a Spaceship or Vindication for a Scout?"

From an ancient Chaldean manuscript was a strange report on "the Marvid, a somewhat odd form of flying craft." The text is quoted as follows:

> The Marvid is powered by *three vibrating spheres* which are fixed to its undercarriage, the angles of their positions must be stable and the pulsations of all three must be harmonised.
> The Marvid is controlled by a graphite rod which brushes the two rear spheres and a copper coil which can be made to surround the front one. Both of these are controlled by a wooden wheel in the centre of the vehicle, which is geared to both of them. The correct setting of the control wheel is indicated by a carefully calibrated crystal which has to be replaced after every journey. The functioning of the aircraft depends entirely on the correct calibration of the crystal.
> The carbon rod was said to have acted in the manner of a carbon brush in a dynamo, while the copper coil increased or decreased power according to its position in relation to the front sphere.

T. Henry Moray's Free Energy Machine

Before the first World War, Dr. T. Henry Moray of Salt Lake City, USA, researched and built a device which produced a thousand watts of power (1.34 hp) for a pound of electrical apparatus or .75 lbs. per horsepower, which is only now realisable with modern, extremely expensive, gas turbine engines. In a description of how his machine worked he said:

> *Oscillations by synchronization* are started in the first stage of the circuit of the device by exciting it with an external power source such as the difference of potential between two points. The circuit is then balanced through synchronization until the oscillations are sustained by *harmonic coupling with the energies of the universe*. The reinforcing action of the harmonic coupling increases the amplitude of the oscillations until the peak pulses *spill* over into the next stage through special detectors or valves which then prevent the return or feedback of the energy from the preceding stage, which oscillate at a controlled frequency and which again are reinforced by harmonic coupling with the ever present energies of the cosmos. That is, the first stage drives a second stage, the second stage drives the third and so on. Additional stages are coupled on until a suitable power level at a usable frequency, voltage and amperage is obtained

by means of special resonant oscillators. Once the device is in operation and delivering energy, it does not require the continuance of the original excitation induced by the difference of potential between two points to maintain the oscillations. *The oscillations are sustained as long as the circuit is completed through a suitable load.*

In the 1930s, the federal government and large business concerns were showing interest in Moray's device and President Roosevelt ordered engineers from the Rural Electrification Administration to work with the inventor. It soon became obvious to Dr. Moray that the government engineers "were not working for the same side" and he charged the REA with attempting to sabotage his work. Then, a federal engineer named Felix Frazer *went berserk* and smashed the device and the equipment with a sledgehammer and with it, Dr. Moray's twenty years of research and development and his entire fortune.

Moray was never able to recover sufficient funds to rebuild it and he died a very disillusioned and embittered man, to be added to the list. At the last hearing his son headed the Research Institute Inc. of Salt Lake City, trying to raise funds to rebuild that generator from his father's notes Fig. 3.A1.

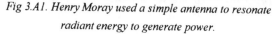
Fig 3.A1. Henry Moray used a simple antenna to resonate radiant energy to generate power.

The Verschoyle 'Aeromotor'

In 1936 W. D. Verschoyle built and demonstrated an electrically operated device which apparently overcame gravity. At that time a film was shown in the principal London cinemas of it lifting a small load to the ceiling of the inventor's workshop. Scientists' naive attempts to explain the phenomenon away as a purely electrostatic effect were easily dispelled by first term considerations and are of little consequence. Verschoyle wrote several guarded articles on his invention before vanishing into obscurity with the onset of WWII.

After the war, attempts to locate him at his last known address in London proved futile. It was believed that he could have been killed there, as that part of London was very heavily bombed. Verschoyle was known to have taken out patents on his device, but all knowledge of the existence of this has been denied. The inventor, together with his device and notes, vanished as if they had never existed.

The Free Energy Device of Lester Hendershot

An American inventor, Lester Hendershot had worked for many years developing a device which eventually generated 140 volts at 60 cycles in the form of a DC pulse. Other scientists worked with him and testified to that. The frustrating thing about his machine, however, was that *it only operated when Hendershot was in near proximity*, even though fraud was completely eliminated by investigators. The reader should remember Keely also had this trouble earlier in his work, but eventually isolated the possible psychokinetic effect completely.

In April 1961, Hendershot agreed to reveal the details of both the construction and operation of his machine in an environment which would include scientists and necessary test equipment to evaluate the nature of the power generated by it. Then on April 26, 1961—less than two weeks after his announcement—Lester Hendershot had died suddenly and mysteriously.

The Electrostatic Motor of Edwin Gray

In the early 1960s, Edwin Gray of Los Angeles produced a fuelless, magnet-driven, "pulsed capacitor discharge electric engine," U.S. patent No. 3,890,548.

For his breakthrough work, Gray was nominated for the "Inventor of the Year" award by the L.A. Patent Attorneys Association. His motor was publicly endorsed by Dr. N. Chalfin and Dr. Gene Wester of Cal. Tech. Dr. Chalfin said: "There is no motor like this in the world. Ordinary electric motors use continuous current and constantly drain power. In this system, energy is used only during a small fraction of a millisecond. Energy not used is returned to an accessory battery for re-use. It is cool running; there is no loss of energy in the system."

Then suddenly, in July 1974, the L.A. District Attorney's Office, apparently acting on orders from 'the top,' raided Gray's plant and confiscated his prototype motor, plans, and records and brought extraordinary transparent charges against him, defying all attempts by Gray's lawyers to retrieve the confiscated materials. Thus by keeping all his funds tied up in legal expenses and preventing further production of his motor, Edwin Gray had been driven into bankruptcy. Which of course is less dramatic than the preceding cases, but equally effective nonetheless Fig. 4.A 1.

Fig 4.A1. Ed Gray demonstrates his 'electrostatic motor'. He is said to have spent $2 million in developing his invention only to have it destroyed by 'economic and political forces'.

The Orgone Energy of Wilhelm Reich

Between 1924 and 1954, a brilliant original thinker, Dr. Wilhelm Reich, claimed to have isolated a new kind of space energy which he called 'Orgone.' During the latter part of his life, Reich successfully developed a method of collecting and directing this energy, which he considered permeates space and all living things—which sounds very similar to the 'ether'—and although no doubt he was aware of the possibility, his interests leaned more towards the pathological associations rather than the technological ones. Much of Reich's work centred on the effects of Orgone on the way we live, healing, growth of plants, etc., but apparently some found even this close brush with Orgone rather sensitive information, demanding that its discoverer should admit his claims as fraudulent, which of course he never did and apparently suffered a term of imprisonment in the United States of America for it.

Reich continued to suffer intense harassment to a degree which sounds not unlike the inquisition, but he never retracted his beliefs.

The Energy Machine of Joseph Newman

One of the most recent and intriguing reports to arrive from America has been that concerning the work of Joseph Newman, a brilliant inventor from Lucedale, Mississippi, who has spent the last two decades working on a machine which he claims will give out more energy than is put in, "to charge the magnets to produce the electromagnetic fields." He claims this is the result of his design "which releases atomic energy in the electromagnetic force field and will provide cheap, clean energy for homes, cars, etc."

Extraordinarily enough, Newman has been engaged in a seven-year controversy over whether he will be granted a patent or not. So far, the U.S. Patent and Trademark Office (PTO) has vigorously refused to grant a patent on the grounds that "it smacks of a perpetual motion machine," even though Newman has never called his device that.

According to *Science News*, Jacob Ribinow, a consultant to the Office of Energy-Related Inventions at the National Bureau of Standards (NBS), submitted a sworn statement in 1984 in which he declared that Newman's machine could not work—without having seen or tested it! It seems the Patent Office had asked the NBS to test a working model and in July NBS released a report stating that its researchers had found the machine's efficiency to "vary from 26% to 67%" but not approaching the efficiency Newman claimed for it. However, a trial was scheduled for December 8, 1986, and the *Science News* article concluded that "the legal battle is likely to end up in the Supreme Court."

On June 7, 1986, some investigators had occasion to talk with a retired government physicist who said, "The machine will work, but they'll never permit it to become operational because it would mean the collapse of the economy." (Shades of John Keely.) It was stated that seven Congressmen had initially introduced a bill calling on Congress to grant Newman a "Pioneering patent."

The number has now risen to eleven.

One investigator quoted a letter written by a former private industry patent examiner to the U.S. Patent Office. The letter continued, "When the U.S. public becomes aware of the scientific situation in which Denmark, Germany, Switzerland and even Spain and South Africa are ahead of us in this field, I think you would do well to reconsider the advancements now being made in this field."

Newman is said to have studied how a magnet is made and theorised that "If the energy that can be produced by a permanent magnet is greater than the energy used to produce the magnetic force field, that energy must come from the atoms in the magnet."

Other Energy Claims

In Germany, the Association of Gravity Field Energy, headed by Dr. Hans Nieper, Hanover, has claims for a "single bearing disc electrical generator" which will last for over ten years, giving out practically free energy. And a "*Cosmic* Energy Generator" comes from a European magnetics expert with 120 patents to his credit, who recently retired from Siemens, a large German electrical equipment manufacturer.

The nature of the foregoing reports suggests that most claimants are working primarily on alternative energy producing machines, in that they presumably supply electrical or mechanical power to vehicles by ordinary means, i.e., wheels or propeller, etc., and although in some respects there is an implied degree of coherence to the present author's work—with the exception of Keely—none of these claim a *direct* implementation of Comat power as a true space propulsion drive or anti-gravity process. In other words, although it is quite possible that such research will produce electrical energy, it is most unlikely to produce direct gravitational or EM drive.

---

[1] Note: This effect, now well known in modern physics as *sonoluminescence*, is a comparatively recent phenomenon (a century *after* Keely) in which electrons are said to change their energy levels due to ultrasonic vibrations, thereby emitting quanta of light.

[2] Ufologists will recognise the significance of such anti-gravity related phenomena.

APPENDIX 2

This giant sized beam engine was installed in 1860 at a hospital in Haywards Heath to drive water supply pumps until 1928. Its average output was a mere 3 bhp. This picture would have to be enlarged by no less than 18 times to be in the scale with the baby sized modern counterpart below.

Shown here actual size. This popular modern IC engine is mass produced and sold world-wide. True its working life cannot match the above beam engine but its replacement costs are proportionally very low.

Comparison of old and present engine sizes.

Meadow grass in centrifuge test cell shows immediate response
to applied negative g.

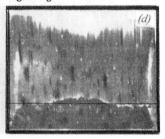

A comparison of (b) and (c) shows the moment in time just before a 'crater'
is formed . The lineal velocity of the cell while these pictures were taken was
approximately 20 feet per second. In these experiments centrifugal force as
a uniform field was used to simulate gravitational effects on samples
contained within the cell in the centrifuge. See Chapter 8.

General view of the author's centrifuge showing test cell and camera (e)

Laboratory equipment built from scrap by the author
with which to examine gravitational effects related to
the Inverse Square Law

The Inverse Square Law (see Chapter 5)

In which $A = \dfrac{R_1^2 \times A_1}{R_2^2}$

A = Area in square feet

R_1 = Radius in feet   1 ft.

A_1 = Area in square feet   4 ft.²

R_2 = Radius in feet   2 ft.

Hence $A_2 = \dfrac{1 \times 1 \times 4}{2 \times 2}$ = 1 sq. ft.

Or if we know Area 2 and require to find what Area 1 would be at double the distance, we get:

$$A_1 = \frac{R_2^2 \times A_2}{R_1^2} = \frac{2 \times 2 \times 1}{1 \times 1} = 4 \text{ sq. ft.}$$

Suppose in this example we have an electrostatic field radiating from a source to which a body is attracted by a force of 16 ounces, when it is positioned only two inches from the source. What will be the force on the body if we doubled the distance to four inches?

Well, by simply altering our units from Area to Force we get:

$$F_2 = \frac{R_1^2 \times F_1}{R_2^2} = \frac{2 \times 2 \times 16}{4 \times 4} = 4 \text{ ounces.}$$

A TIME FOR EVERYTHING

The following few typical extracts from my book *The AT Factor* have been included in this Appendix in support of the precognitive tendencies in their content. Of course, standing alone as they consequently are, this value is accordingly reduced. But this can be rectified somewhat by correlating them to some of the other instances I have related in earlier chapters, bearing in mind the fact that I have filled an entire book with more of them. It must also be stressed, I am not so much interested about the fact that I took part in these incidences, so much as that they happened at all. The abbreviation ATF (Advanced Time Factor) simply offers a visual measure—as near as possible—of the time relationship.

There are, of course, other connotations and circumstances involved which surround these technological excursions, a further discourse on which would not be appropriate here. Suffice to advise the reader to accept that any precognitive tendency *should not be excluded from the main directive of this book.*

However, it may not be generally known that particularly in the aerospace industry, the state of the art technology, when published, can be anything up to ten years old. In other words, it can be taken as a certainty that in some parts of the world there is work currently in progress on advanced technology which is so secret as to render its general acquisition well into the next decade or so, depending of course on the degree of its advanced status.

In this sense, the publication of some of the author's advanced designs in *The AT Factor* is, in effect, a peep behind the secret status curtain. However, it would be correct to interpret such an otherwise overgenerous action as being a measure of the totally encompassing nature of the technology portrayed in *The Cosmic Matrix*.

1944 - 1955 SURFACE EFFECT MODELS
AND PLANS TO HELP MEET THE BILLS
ATF 11 Years

This work preceded the investigation by Sir Christopher Cockerell by over a decade. Extracts are from a series of articles including the Cockerell look-alike model which was *exactly the same size and colour*. It caused bewildered amazement at one of the world's largest test tanks in the south of England. Among other things it was this work which heralded the beginning of hovercraft skirts.

There were another two incidents of precognitive nature. On an occasion when the author's brother, noticing the 'Hoverer' model cutting a long swath in the grass in *1950*, said, *"Wouldn't that make a smashing lawn mower?"* (which considering the number of *hover* mowers there are now all over the world . . .!). Then later—having made a successful demonstration on the flat roof of a hotel in Southampton to representatives of a shipbuilding firm from the Clyde in Scotland—the fan disintegrated. The author looked up in time to see a naval architect tucking a small fan blade into his waistcoat pocket. Responding sheepishly the man said *"You never know, there might be a little piece of history here!"*

Model Hover Craft

An account of experiments
including a practical model

By Leonard G. Cramp, A.R.Ae.S., M.S.I.A.

THE increasing public interest in hovercraft type vehicles has encouraged the writer to offer this brief account of some work carried out a number of years ago.

It will be seen that the original ideas pursued then were, in many respects, identical with present trends of development, moreover it must be stressed that this research was of a private nature and does not intentionally represent existing designs.

The principle of hovercraft or air cushion borne vehicles is now generally understood, but a brief word or two as an introduction will not be amiss. It is basically a means of differentiating air pressure above and below a pressure plane, but—unlike an aeroplane wing—there is no reduction of atmospheric pressure above the plane to any marked degree. Fig. 1A shows such a plane where it will be seen the supporting surface or ground is in

explosions. Such a device is subject to excessive vibration, but it served to prove the point.

In operation it is similar in principle to the so-called plenum chamber type craft now being investigated in Great Britain, Switzerland and the United States of America, Fig. 1d.

The writer's conception of a forced air operated machine on these lines was approached from a slightly different aspect than that of a pure plenum chamber; the more developed theory is shown in Fig. 2a. Air was passed through a propeller as in Fig. 2a. Air was passed through a propeller as in the plenum chamber Fig. 1d, escaping to atmosphere via the periphery as in the previous case. But the underside of the craft was so designed as to direct the airflow smoothly downwards towards the supporting surface where it curved away towards the rim of the craft. Thus it will be seen from Fig. 2a that there is a relatively large mass of trapped air

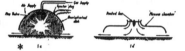

Fig 1a   1b   * 1c   1d

fact one of two restraining walls trying to prevent the incoming air from escaping, while the second wall is formed by the plane itself.

It will be appreciated that the aperture through which the expanding air escapes has a critical function, its area increasing constantly with increase in pressure plane size. Theoretical available work is completed when the air pressure is expanded down to the surrounding atmospheric pressure at the exit.

The natural development of the idea is to form the plane with a skirt as in Fig. 1b, in which it will be seen that some of the kinetic energy in the outflowing stream is again converted into pressure energy through turbulence.

A conducted experiment embodying this basic principle is illustrated in Fig. 1c. In this the pressure plane was formed by an inverted dish shaped bowl into which provision was made for a constant supply of coal gas. A simple flap valve regulated air induction, whilst a small glow plug was employed for ignition.

In effect, the whole formed a simple combustion chamber, in which a mixture of gas and air could be ignited. The operation is almost self explanatory. First the igniter was lit, then gas was allowed to flow into the chamber. As soon as a detonating mixture was reached, the charge ignited, the resulting increase in pressure forcing the dish upwards. The instant the dish left the supporting table, the gases escaped, the ensuing suction bringing in a fresh charge, and the process continued.

In effect it will be seen the principle was identical to the pulse jet cycle employed in the VI during the war.

By this means the dish was kept "airborne" by the increase in pressure caused by the high frequency

between the pressure plane and the outgoing stream and it will be apparent that due to viscosity this mass of air will be induced to revolve in the fashion of a toroidal belt, its velocity approaching V—that of the outgoing stream. In practice this was found to have been in the order of 90 per cent. of V and therefore represented quite a usable amount of energy.

The manner in which this energy is converted into lift is quite simple and calculations were based on the following basic assumptions. The revolving mass of air exerts an upward thrust on the underside of the pressure plane due to centrifugal force, whilst the outward moving stream suffers a downward but equal centrifugal "pinch"

Model above is as shown on G.A. opposite. Below is the top view after modification

---

* This experiment was first conducted at Hatfield, Hertfordshire, in 1944 and the resulting 'Hoverer' model bore such a remarkable resemblance - including finishing colour - to the model built for Sir Christopher Cockerell, that it caused confusion at the Saunders Roe, Isle of Wight, test tanks when his model was later tested.

1942 - 1994 MINIATURE GAS TURBINE EXPERIMENTS
ATF 52 Years

The following shows extracts from a series of articles in 1948 concerning the author's work on a small gas turbojet engine. It is interesting to compare this with the modern counterpart only now being developed. The De Havilland Vampire scale model aircraft was built by the author to suit this engine, having been scaled from the original general arrangement of the experimental Vampire prototype (recognisable by the economy type *square cut tail fins*). All other versions had the distinctive De Havilland *elliptical* profile fins. Yet, not only are such small turbojets only now becoming available, but one of the *first* has been installed in a model *Vampire*, which as the following photographs show, it too, has the experimental *square cut fins*!

GAS TURBINE EXPERIMENTS

PART I.   By L. G. CRAMP.

ONE seems to hear little or nothing nowadays on the subject of model jet propulsion, and the writer having carried out some research in this field over the last four years, would like to contribute a little toward reviving some interest among modellers.

No doubt among the readers of the AEROMODELLER there are those who will sympathise with the writer, who is often painfully reminded of the fact that the airscrew power driven

This was found to be quite successful. The chamber was heated and while the preliminary fuel was still burning, the fuel valve was opened a little and at the same time the cycle crank was started up. The little fan screamed around at something like 25,000 r.p.m. and the unit gave forth that very pleasing blow torch roar, emitting a flame several feet in length. This became blue in colour and shorter in length as the fuel valve was operated, permitting a more correct air-

Flame tube

Flame tube

Gas Turbines 40 years ago??

Reproduction of a heading to a reader's item from the Winter 1993 issue of the popular Aeromodelling Magazine ,'Jet International'.

The modern realisation (centre) of the small turbojet engine compared with the author's version of 1942 in which even the flame tubes are identical.

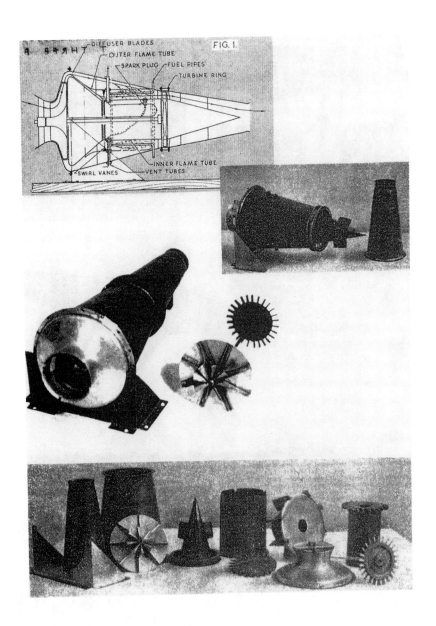

FIG. 1.

DIFFUSER BLADES
OUTER FLAME TUBE
SPARK PLUG
FUEL PIPES
TURBINE RING
INNER FLAME TUBE
SWIRL VANES
VENT TUBES

When the turbine was run up at De Havillands in Hatfield UK, staff mistook the din for the full size 'Goblin' jet which powered the Vampire aircraft.

Author's son Gary then only 8 years at the time.

The very first free flight scale model of the De.H.Vampire, designed and built by the author in 1947, from the makers original plans, which caused bewildered consternation among onlookers at a UK coastal holiday resort in 1983. Note the prototype square cut fins. Compare with recent model below.

Pete Marsden, the first pilot to fly a JPX powered model in the UK, has recently built this own-design DH Vampire as a testbed for a future more 'scale' version. The model is 76" wingspan, weighs only about 13Lbs and flies like a 'trainer'. Watch out Pete's article on the new scale version in a future RCJI, together with some clever weight saving construction techniques.

1956 - 1985 FROM SMALL BEGINNINGS
THE COLD TURBINE RAM-JET
ATF 29 Years

The work on the small turbojet was directly responsible for the Cold Turbine Ram-Jet in 1956. Although the more recent look-alikes are practically identical in basic concept, there is no known reason to doubt that these latter versions are completely original.

A cold turbine ejector engine

A new form of gas turbine engine, in which the turbine is relieved of high gas temperatures, is proposed in this article. Protective rights on this, the author's own idea, have been filed by a major aero engine manufacturer

In the aeronautical world it is natural for the word turbine to be associated with thoughts of intense heat, and we accept the fact that it is chiefly the turbine and the stator ring in a turbojet engine which takes the full blast of the hot expanding gases.

The flame temperature of a modern gas turbine can be in the order of 2 000 degrees C, and the gases have to be 'diluted' and cooled down to a temperature of about 850 degrees C before they pass through the stator and turbine. This is known as the 'working fluid temperature' (w f t) and upon its magnitude largely depends the efficiency of the cycle.

The w f t is usually governed by the temperature at which the turbine can operate, and when it is borne in mind that the turbine in a modern turbine engine runs red hot—suffering extremes of centrifugal and bending loads in the process—some of the problems which have beset the metallurgist can be appreciated. A turbine blade which, at rest, weighs only a few grammes, can experience a disrupting stress of something like 15 kg mm² (10 ton in²) due to centrifugal force alone.

Naturally, under these conditions the blade is inclined to elongate which, if unchecked, can—and does—lead to disastrous consequences. Largely due to the enormous strides made in metallurgy, which has furnished gas turbine designers with better materials, the permissible w f t has been pushed higher and higher, and for each small increment of temperature, so has the power output of engines soared. If the inlet temperature to the turbine can be raised, not only will the thermal efficiency be increased, but the quantity of air required for cooling the combustion products will be reduced progressively, so that a smaller compressor and turbine can be used for the same net power output (see Figure 1). It is to this end that designers are constantly striving.

One of the most promising solutions to the problem currently receiving attention is to form hollow turbine blades into which is fed cooling air. This, unfortunately, can only be obtained by further complexity, which, of course, incurs weight, manufacturer's costs and maintenance penalties.

In this arrangement air is 'bled' off the main compressor staging and is induced to flow through various intricate labyrinths and passages to the main compressor turbine rotor shaft, and then to the turbine disk itself. Further passages in the disk convey the air through the roots of the hollow turbine blades, where—now considerably heated —it passes through small outlet holes situated in the blade trailing edge, as in Figure 2.

Research and development work on these lines is going on throughout the world, while the metallurgist continues to find stronger and greater heat-resisting alloys.

From all this it will be appreciated that should it be at all possible to extend the w f t any higher, even greater power could be attained.

L G Cramp

The following brief report on the writer's attempts to find a solution to the problem may suggest a possible way of achieving maximum w f t, though it is pointed out that the principle outlined here requires further examination. Early enquiry into the cycle's functional capabilities has indicated that it would work, though it might have limitations which need not be prohibitive.

It is now well known that in the conventional gas turbine cycle the pressure energy contained in the hot gases is converted into kinetic energy in the form of an expanding jet of high velocity gas, which impinges on the turbine blades giving up some of its energy to the rotor system and releasing the remainder through the propelling j... for jet reaction.

The 'ejector' engine, as the writer's engine is called, is somewhat different in operation, deriving its name from the ejector principle. As with the ramjet it offers the advantage that gas dilution is unnecessary, inasmuch as the products of combustion never come into contact with the turbine at all. Instead, provision is made for the rapidly expanding high velocity jet to move a large volume of cold air, in which is placed an air turbine.

The principle is basically simple and automatically suggests the engine's component layout, one of which is shown diagrammatically in Figure 3c.

The original conception (Figure 3b) may serve to convey the natural development of layout. In this configuration we have a conventional turbine-compressor assembly as in Figure 3a, but with the exception that, at the nozzle box, where the hot gases enter the stator blade ring, there is placed a duct le... to atmosphere. Although it will be appreciated that this is an unworkable arrangement, it may serve to convey the general idea in its broad sense.

It will be apparent that the high velocity gas jet will induce an inflow of cold air, giving up some of its temperature to the air in the process. Therefore we could expect to be able to increase the w f t. Indeed, this would be necessary, for the hot gases would also have given up some energy. In addition, the turbine would still be exposed to heat, so there would be no obvious gain in this direction.

But the cold air entering the duct has considerable energy, energy which was given to it by the expanding gases, which would normally drive the turbine direct. Theoretically—therefore—allowing for losses—the volume of cold air should have almost enough energy to drive the turbine compressor set. And in any case we can now increase the w f t to a maximum.

Re-arrangement of the components

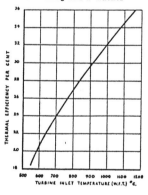

Figure 1—This graph indicates the improved overall efficiency of the cycle which occurs when the temperature of the working fluid is increased

THERMAL EFFICIENCY PER CENT

TURBINE INLET TEMPERATURE (W.F.T.) °C.

By 1960 the author had moved to the Isle of Wight and due to the shelving of the cold turbine project he took the step of publishing this article on it.

gives the layout shown in Figure 3c. From this it will be seen that the turbine and compressor are integral, so arranged as to form two-tier blading on a common rotor disk. The tiers are separated by a ring which separates the blades and also forms the continuation of the turbine-compressor intake ducting, Figure 4.

Figure 3a shows, for comparison, a conventional jet propulsion gas turbine layout of similar compressor and turbine staging, while the structural and general advantages, which chiefly affect size and weight, are as follows :—

(1) The overall length of the engine is greatly reduced.

(2) The costly and heavy rotor shaft is eliminated.

(3) The turbine rotor, being integral with the last stage of the compressor assembly, represents a further saving in manufacturing and assembly costs.

(4) Flame length, which seriously influences overall size in the conventional turbine engine, is of little or no importance.

Although these represent the more obvious advantages, closer consideration will reveal others of not so dramatic a nature, but which are enhancing, nevertheless.

The starting procedure of the ejector type engine would be similar to present-day turbojets. The rotor is motored up to operating speed, drawing in air which suffers a rise in pressure as it passes through the compressor stages. Passing into the diffuser accompanied by further pressure increase it is delivered into the

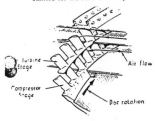

Figure 4—This illustration shows the arrangement of the two-tier blading. The guide vanes and stator rings are omitted for the sake of clarity

combustion chamber as the primary airflow.

Fuel is then burned in the normal way and the ensuing products of combustion are expanded at high velocity through a convergent type nozzle into the mixing zone.

The considerably large volume of air thus expelled from the rear ' mixing ' chamber causes a pressure drop in the air turbine annulus, which in turn induces a high velocity flow of cold air across the turbine. The energy imparted by this secondary airflow to the turbine is absorbed by further work done in the compressor and the cycle is then completed.

Effectively then, the energy in the

products of combustion are absorbed by the turbine as in normal practice, except that part of this energy is employed to move a protective belt of cold air which, in turn, operates the turbine. As already stated, this represents, to some extent, a drop in the efficiency of the cycle, but it must be remembered, on the other hand, that the w f t can now be maintained at a maximum, requiring no further dilution.

A hypothetical engine has been designed to meet these conditions, and all but one stage of the axial compressor has been eliminated. This, in effect, reduces the moving parts of the engine to little more than one two-tier rotor, the inner ring comprising the compressor stage, and the outer, the turbine. A similar arrangement to this was employed on several ducted-fan augmented jet engines.

In Figure 5, normal stator rings are included for both turbine and compressor, and it was anticipated that variable inlet guide vanes would be necessary.

The writer—having successfully carried out experiments with small centrifugal type fuel injectors—is of the opinion that this type of pump would suit the basic simplicity of the ejector engine ; the remainder of the entire unit could be fabricated.

Should it be proven that the engine cycle would not sustain itself under static conditions, that is, it would require some ram effect from forward motion, then perhaps the addition of a small liquid fuel rocket might prove an attractive alternative, as illustrated in Figure 5.

Such an arrangement should prove to be quite a space-saver in mixed unit type aircraft. In this case, the cycle of operations would be for the rocket to be fired and the rotor ' motored ' up to operating speed simultaneously ; the resulting increased mass flow through the engine being employed not only to accelerate the vehicle to its critical

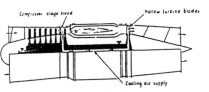

Figure 2—Cooling air is bled off the compressor and led to hollow turbine blades

Figure 3—The three drawings below indicate the natural development of the ejector engine

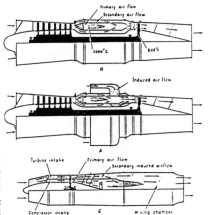

ram speed, but to offer further increased thrust for take-off.

At the time of writing there is news of a new American turbo-ramjet propulsion unit, and although this title would seem to convey a similar idea to the ejector engine it is possible that it differs in its basic concept.

* Results of a preliminary design study by D Napier & Son Limited revealed that the economic operational margin might be restricted and, in fact, it was shown that the simple cycle would not be self sustaining under static conditions. However, the study also revealed that the ejector type engine would yield a thrust co-efficient and a specific fuel consumption comparable with a ramjet at transonic speeds. Protective rights for the ejector engine have been filed by D Napier & Son Limited.

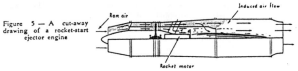

Figure 5 — A cut-away drawing of a rocket-start ejector engine

⁓⁁TENT SPECIFICATION

DRAWINGS ATTACHED
Inventor: LEONARD GEORGE CRAMP

Date of filing Complete Specification Nov. 1, 1957.

Application Date Sept. 3, 1956.

Complete Specification Published Feb. 10, 1960.

827,663

No. 2(922/56.

PATENTS
DEPARTM NT

D. NAPIER & SON
LIMITED
V 3.

Index at acceptance: —**Class 110(3),** G(1A : 19), J1A, J2(A1A : A1C : B4 : BX).
International Classification: —**F02c, k.**

COMPLETE SPECIFICATION

Internal Combustion Engines

We, D. NAPIER & SON LIMITED, a company registered under the Laws of Great Britain, of 211, Acton Vale, London, W.3, do hereby declare the invention for which we pray that
5 a patent may be granted to us, and the method by which it is to be performed, to be particularly described in and by the following statement: —

This invention relates to internal combus-
10 tion engines and is primarily applicable to aircraft propulsion units.

An internal combustion engine of the continuous combustion type according to the invention comprises an air compressor
15 arranged to deliver air to a combustion chamber, and means for delivering fuel to be burnt with the air in the combustion chamber, the hot combustion products being allowed to expand and discharged at high velocity
20 through an exhaust aperture, the compressor being driven solely by an air turbine arranged in a relatively cool air stream created or assisted by the high velocity gas flow discharged from the combustion chamber.
25 Thus an important advantage of the present invention over a conventional gas turbine engine where the air compressor is driven by a gas turbine arranged in the hot exhaust stream, is that the turbine is subject only to
30 the relatively low temperatures of the air stream.

According to a preferred feature of the invention the air turbine lies in an air duct which communicates at its downstream end
35 with a common exhaust duct into which the hot high velocity gas flow from the exhaust aperture also passes, the junction between the two ducts being in the form of an ejector such that the high velocity gas flow
40 induces or assists the flow of air through the air duct.

Where the engine is mounted on an aircraft to act as a propulsion unit the air flow through the air duct will of course be ampli-

fied by the "ram" effect of increased air 45 pressure arising from the forward velocity. In fact it is thought that some "ram" effect will be essential for the cycle of the engine to be self sustaining.

According to another preferred feature of 50 the invention the compressor and turbine are both of the axial flow type and are mounted on a single rotor assembly in the form of radially spaced tiers or rings of blades. This simplifies the construction of 55 the rotor assembly and the air duct is then conventionally arranged as an annular passage surrounding the combustion chamber and communicating with a common exhaust duct at the downstream end of the chamber. 60

Means may also be provided for supplying fuel and burning it in the gas flow downstream of the exhaust aperture.

The invention may be performed in various different ways but one specific embodiment 65 will now be described by way of example as applied to an aircraft propulsion unit of the jet propulsion type, illustrating in section in the accompanying drawing (which is partly broken away and shortened for convenience). 70

In this example the engine comprises an outer elongated generally cylindrical casing 10 which is tapered slightly inwards towards its upstream end 11 (which constitutes an air intake) and within which is mounted an 75 inner cylindrical casing 12 extending lengthwise over a limited part of the total length of the outer casing and spaced appreciably from its two ends. The inner casing is supported from the outer casing by three series 80 of stream-lined spats or struts 13, 14, 15 to provide an outer annular air duct 16, surrounding the inner casing.

Adjacent the upstream end of the inner casing 12 a central bearing assembly 17 en- 85 closed within a bullet-shaped fairing 18 is supported on inner extensions of the radial spats 13, 14 and in this bearing assembly

[Price 3s. 6d.]

Copy of D. Napier's original patent application.. Napier's technical
writeup vanished from the author's file.

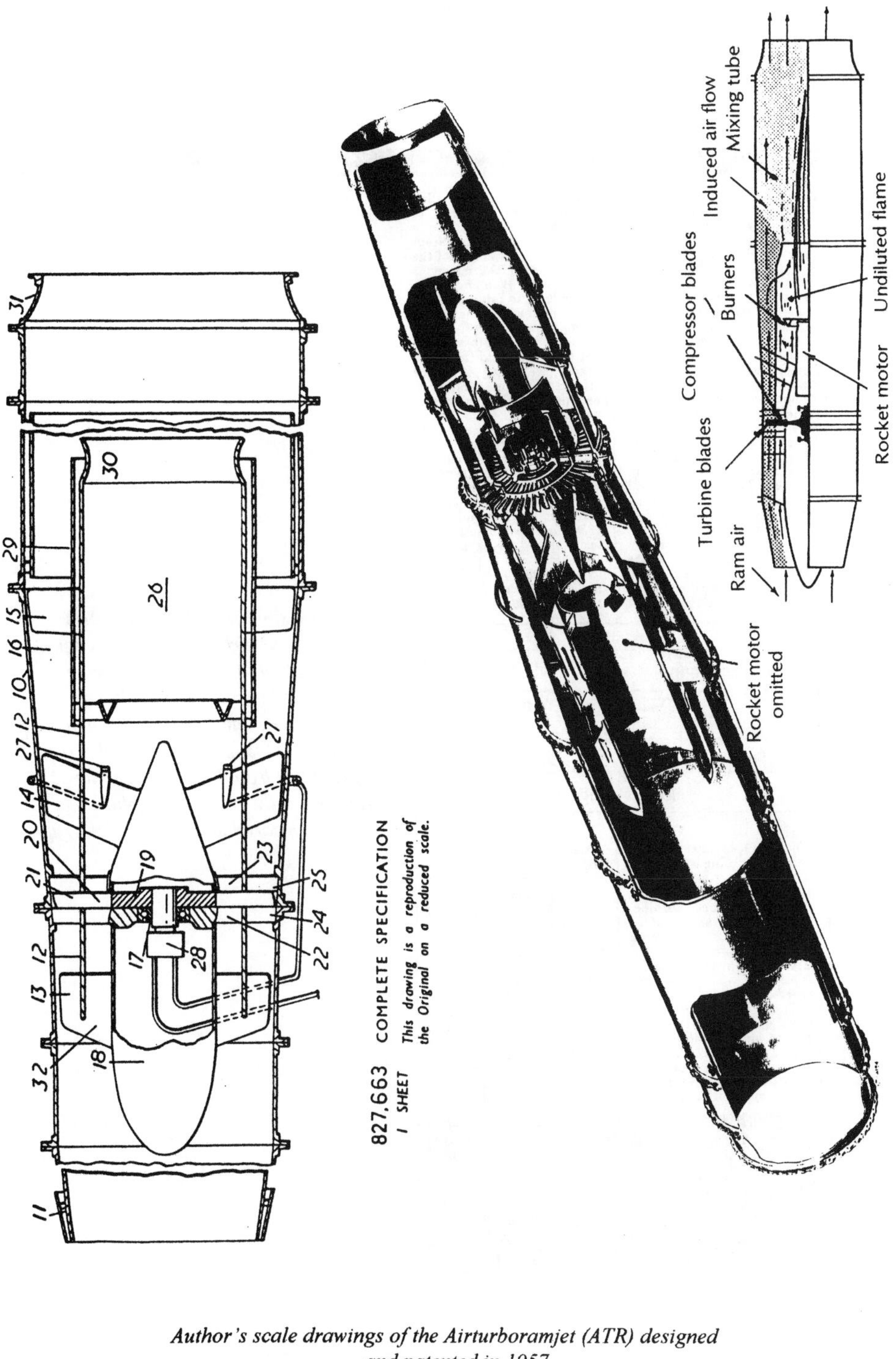
827,663 COMPLETE SPECIFICATION
1 SHEET
This drawing is a reproduction of the Original on a reduced scale.
Induced air flow
Mixing tube
Compressor blades
Burners
Undiluted flame
Turbine blades
Rocket motor
Ram air
Rocket motor omitted

Author's scale drawings of the Airturboramjet (ATR) designed and patented in 1957.

Continued from page 16
Extreme stresses may also be present owing to the speed and weight of the vehicle. For craft travelling at up to Mach 5, titanium or advanced aluminium alloys are likely to be used, probably shaped using superplastic forming techniques.

At greater speeds, carbon-carbon or ceramic matrix composites, and double-skinned metal matrix composites, are among the materials proposed. Again, small components such as Space Shuttle wing leading edges (manufactured from carbon-carbon) are already in service. The technological challenge will lie in adapting laboratory processes for large-scale manufacture.

British Aerospace's Hotol concept, first presented at the 1984 Farnborough Air Show, is very similar to some of the US proposals. At the time of writing, BAe is awaiting around £2million-worth of government funding for a two-year proof-of-concept study into the viability of the spaceplane. If the money is forthcoming, BAe and Rolls-Royce hope to bring Hotol to the point at which it is seen as a programme worth full-scale funding. If the project does not then appeal to the European Space Agency, it may be offered as the basis for partnership in a US programme—particularly if Rolls-Royce's liquid-air-cycle engine concept proves workable.

Darpa's current timescale for a hypersonic transport programme envisages engine/airframe evaluation over the two years beginning in January 1986, construction of a flight-test vehicle taking place in 1988-1991 and flight-testing in 1991-1995. Around $3,000 million will be needed to reach the flight-test stage.

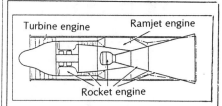

Turbine engine   Ramjet engine

Rocket engine

An airturboramjet (ATR) seen undergoing static test at Aerojet's Sacramento, California, facility. The engine combines turbine, rocket, and ramjet technology, operating like a turbojet subsonically and like a fan-boosted ramjet at supersonic and hypersonic speeds.

The ATR's atmospheric cycle is similar to that of a conventional turbojet. The engine's core contains a compressor to increase the ramjet pressure ratio. This is driven by a turbine, which, instead of being open to the airflow through the engine, is enclosed in the powerhead housing, where it can be driven by a high-pressure fuel-rich hot gas from a separate combustion chamber. Fuel is partially burnt in this chamber, then passed to the nozzle chamber for final combustion. Alternatively, cryogenic fuels can be used to provide active cooling of flight surfaces, becoming vaporised in the process, the vapour being used to drive the turbine before being burnt in the nozzle.

The close similarity to the author's original cold turbine aerospace engine concept of 1960 and that now being investigated in the UK and USA as an 'exciting new idea of breakthrough proportions' is apparent from this article reproduced from Flight International 1985. More recently in 1989 the following article appeared in the Daily Telegraph.

FLIGHT INTERNATIONAL, 7 December 1985

SCIENCE & TECHNOLOGY

Inventor Alan Bond hopes a secret new engine will give extra thrust to his precious space plane project

Could Satan help Hotol take off?

SPACE TRAVEL

Adrian Berry

ALAN BOND, frustrated inventor of the space plane Hotol, is not defeated yet. While the project has collapsed because of the Government's refusal to fund it, he has re-designed a secret engine for the craft. Unlike the earlier design, whose patent he sold to Rolls-Royce, the new one remains his own property, to sell to whom he chooses.

"It's called Satan," he said. "It's significantly more efficient than the original Hotol engine, but I'm not revealing the details to anyone unless I can sell it. I can't even reveal what Satan stands for. There's nothing sinister about the name, but it's an acronym that might provide clues as to how it works. I shall not patent it, since that would risk it being classified. I believe the Government has no legal power to stop me from selling it, even to a foreign government, if that should be necessary.

British Aerospace's design for Hotol are revealed on this page. If Satan gets off the ground, Hotol (short for Horizontal Take-off and Landing) will carry cargo into space at less than a fifth of the cost of the US space shuttle. And for a third of its total construction cost, it could be upgraded to carry 20 to 30 astronauts, three times more

Satan is similar in size to the earlier Hotol engine, so there is no question of having to redesign the overall 275-ton vehicle. Omitting secret details, it will save weight by flying with a 'ramjet' air-breathing engine that will take its fuel from the atmosphere itself.

It will take off horizontally like an aircraft, boosted by an engine component fuelled by liquid hydrogen. It will then climb with its ramjet to a height of some 18 miles (six miles higher than the cruising altitude of Concorde) where it will have reached six times the speed of sound. There, where the atmosphere becomes too thin to provide more fuel, its on-board oxygen would be used to combust the liquid hydrogen, raising the craft to 25,000 mph, the speed necessary—to orbit.

It is far less wasteful than the US Ariane rockets, since, it would land on an airfield, just as it took off. It would land on an airway without jettisoning any of its parts, just like an airliner.

Satan gives it a new dimension. "We haven't yet been briefed about this new engine," said Dr Robert Parkinson, manager of future launch systems at British Aerospace, "but it sounds as if Alan has something good up his sleeve."

British Aerospace, which still hopes to manufacture and market Hotol, has 60 people working on it, and they have just produced the next detailed blueprint of the whole craft. "An improved engine design will eventually make Hotol more viable," a spokesman said.

Bond, now 45, has spent all his life designing rockets, from what were little more than child's toys to the engines of starships. None of them has yet flown. While still a schoolboy, he saved up his 7s 6d a week pocket money for several months to build Poltergeist, a 4ft 6in projectile that he planned to launch from a 19ft tower on the Derbyshire moors. It was going to fly to a height of 10 miles, to beat the American amateur record of the time.

Forbidden to fly it by the Air Ministry because of the Explosives Act of 1875, and threatened with a fine of £100 for every day the rocket remained in his possession, he turned his mind to spaceships of the distant future.

In a team formed by the British Interplanetary Society, he helped to design Daedalus, an unmanned starship to be crewed by intelligent robots that, during the 21st century, may fly to Barnard's Star, six light-years from Earth, in a one-way trip that would last 50 years cruising at 84,000,000 mph, 12 percent of the speed of light.

work by nuclear fusion. It is an even more spectacular idea than Daedalus. With thrust increased by injection of the elements cadmium and thorium into its reactor, it would suck in the hydrogen fuel of interstellar space for its starship engines for hundreds of thousands of cubic miles around the ship, giving it an estimated speed of 70 percent of the speed of light.

His Satan engine is his last attempt to build a successful rocket. "If I can't sell this one, it's my last throw." None of his friends believes this.

He also published in a learned journal a blueprint for Rair, short for Ramjet Augmented Interstellar Rocket, a far-futuristic version of Hotol that would

THE CHANGING SHAPE OF HOTOL

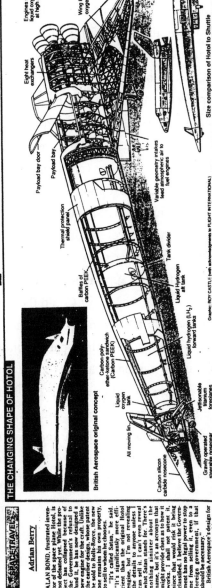

Engines switch to liquid oxygen fuel at high altitude · Wing liquid oxygen tank · Eight heat exchangers · Payload bay door · Payload bay · Thermal protection shield panel · Variable geometry intakes feed atmospheric air to fuel engines · Tank divider · Baffles of carbon PEEK · Liquid Hydrogen (LH₂) aft tank · Carbon-poly-ether-ketone sandwich (Carbon PEEK) · Liquid oxygen tank · Liquid hydrogen (LH₂) forward tanks · Jettisonable titanium foreplanes · All moving fin · Carbon silicon carbide nosecone · Gravity operated steerable nosegear · British Aerospace original concept · Size comparison of Hotol to Shuttle

GRAPHIC: ROY CASTLE (with acknowledgements to FLIGHT INTERNATIONAL)

The proposed 'Hotol' space craft which employs the 'new' ram jet engine. Note the bracketed ram jet description above...

The March of Time

As stated at the beginning of this book, the foregoing items—along with quite a few others—were originally intended for a separate volume in which precognition would have been only one of the contributing themes. I have no difficulty in remaining dispassionate in this context, for I have observed identical phenomena among members of my own family and others. Such 'strangeness factor' is by no means unique, it occurs to many people from all walks of life all over the world, and the tongue in cheek convenient dismissal of "nothing but sheer coincidence" begins to ring more than a trifle tiresomely thin.

It is quite juvenile escapism to continue ignoring these issues, for anyone who seriously devotes time and patience soon discovers a kind of unity. Many years ago and since, I have had these facts thrust upon me, so that now I have conceded to the difficult to accept conclusion that, not only does our species travel 'in time' but the astonishing truth may well be the 'system' incorporates a condition for the furtherance of life which *depends* on it! For if indeed, now and again—as many are apt to believe—mankind appears to be helped or 'inspired' by external agencies, then logically we may ask why? Could the answer be buried in a space-time continuum state which actually requires an advanced species to nurture a lesser? In other words, do 'they' *have* to take part in planting *the seeds of their own future* just as surely as the gardener plants for his? For that matter, as do intelligent humans consider it necessary to teach their children at school in order to perpetuate the species. I suggest that somehow, either knowingly or unknowingly, such a process is endemic. But enough, for that is *quite* another story. .... save perhaps to add that I trust I have satisfactorily answered the oft raised question mentioned at the very beginning of this preamble, *Why another book?*

AUTHOR'S PROFESSIONAL BACKGROUND

Leonard Cramp was born and educated in London and comes from a family of engineers and inventors.

After demobilisation from the Army in 1941, he spent several years surveying and setting out airfields and runways in Norfolk.

The remainder of his working life has been spent in the aerospace industry in England at the following establishments where he was involved in the indicated projects.

De Havilland Aircraft Corp., Hatfield
Mosquito, Hornet, Comet, Vampire aircraft
D. Napier & Sons, London
Forward Projects Dept., Deltic compound engine, the Eland and Oryx gas turbine engines etc.
Saunders Roe Ltd., Isle of Wight
Mixed unit supersonic aircraft SR53, SR177, Scout helicopter, Black Knight/Blue Streak rocket program etc.
British Hovercraft Corp., Isle of Wight
The first SRN1 Hovercraft through SRN2 - SRN6 Forward Projects Dept.
N. D. Norman Aircraft, Isle of Wight
Firecracker aircraft design team with Desmond Norman up to 1977, when Leonard Cramp decided to set up his own company called Airbilt Ltd., to conduct private research on VTOL aircraft, which led to the development of his own concept called the 'Hoverplane,' now under investigation.

BIBLIOGRAPHY

Adamski,G. *Inside the Spaceships.* Arco Publishers. 1956.

Cook,E. *Man, Energy, Society.* W.H.Freeman & Co. 1976.

Davies,P.C.W. *The Search for Gravity Waves.* Cambridge Univ. Press. 1980.

Delgardo,P.& Andrews,C. *Circular Evidence.* Bloomsbury Pub.Ltd. 1989.

Dunne,J.W. *An Experiment with Time.* Faber. 1927.

Forman,J. *The Mask of Time.* Macdonald & James. 1978.

Fuller,J.G. *The Ghost of 20 Megacycles.* Souvenir Press. 1985.

Good,T. *Beyond Top Secret.* Sidgwick & Jackson. 1996.

   Alien Base. Century. 1998.

Gribbin,J. *Time Warps.* J.M.Dent & Sons Ltd. 1979.

Gribbin,J.& Reese,M. *Cosmic Coincidences.* Black Swan. 1983.

Hawking,S. *A Brief History of Time.* Bantam Press. 1996.

Hill,P.R. *Unconventional Flying Objects.* Hampton Roads. 1995.

Hillary,Sir.E. *Ecology 2000.* Michael Joseph. 1984.

Hoyle,Sir.F. *The Intelligent Universe.* Michael Joseph. 1984.

Leroy,O. *Levitations.* Burns, Oates. 1928.

Lethbridge,T.C. *The Power of the Pendulum.* Penguin. 1997.

McCampbell,J.M. *Ufology.* Jaymac-Hollman. 1973.

Michell,J.&Rickard,R. *Phenomena,A Book of Wonders.* Thames & Hudson . 1977.

Paijmans,T. *Free Energy Pioneer,John Worrel Keely.* Illuminet Press. 1998.

Pedler,K. *Mind over Matter.* Eyre Methuen. 1981.

Pond,D. *Keely's Secrets, 'Universal Laws'.*The Message Company. 1990.

Ross,D. *Energy from the Waves.* Pergamon Press. 1981.

Scorer,R.S. *Pollution in the Air.* Routledge & Kegan Paul. 1973.

Walters,E & F. *The Gulf Breeze Sightings.* William Morrow & Co. 1990.

Watson,L. *Super Nature.* Sceptre. 1997.

INDEX

Acceleration, 127
  forces, 235
accelerator, Fermi, 43
acid rain, 41, 74
acronym Syncomat, 25
Adamski, George, 281, 282
Adamski mother ship, 277, 283
advanced technology, 340
  time factor, 340
aerodynamic drag, absence of 285
aerospace, 319
aircraft lift propulsion, 271
air gap syncomat units, 271
Aldrin, Buzz, 74
alternative energy, 39
angular momentum, 279
anti-gravity, 184, 244, 247, 277
  systems, 278, 323
apparitions, 159, 168
  examples, 159
archimedian buoyancy, 89, 241
area cube law, 285
Area 51, 24
Armstrong, Neil, 74
asteroid errant, 306
atmospheric pressure caused by gravity, 93
  drag, 241
  lift component, 251
  magnetic analogy, 92, 93
  pressure, 92, 93, 262
atom formation analogy, 101, 102
attraction and repulsion unified, 228-230
auric field, 159
auto levitation, 177-183
  control systems, 243
Avenel, Antony, 25, 26, 37

Balloons, 85, 89, 302
  hot air, 217
  opposes gravity, 217
beam steam engine, 297, 337
big bang cosmological theory, 129
  unification of, 129
bioplasmic extension, 173
  detector, 203 - 205
biorythmic crystals, 212
bird ingestion, no risk, 319
Birkness, Prof.B.J. experiment, 221
black box, 22, 302

Bondi, Herman, 129
Bopal, 72
braking system, 315
Briggs, J.B. 106
British Book Centre, 11
  Broadcasting Corporation, 47
  Interplanetary Society, 11, 56
Broglie de, 90
bumble bee syndrome,314

Catalystic reaction on modulation, 183
Catoe, Lynn E, 21
centrifugal fans, 131
  disintegration, 300
  force, 87, 98
centrifuge tests, 338
centuries before its time, 329
change, total, 305
Churchill, Sir Winston, 12
circuit of creative rays, 113
clairaudience, 158, 166
Clarke, Arthur C, 11, 39, 40
  2001, 315
Cockerell, Sir Christopher, 60, 70, 341
collective unconcious, 156
comat, all matter subservient to, 72, 231
  alternative nature of, 234
  energy, 46, 89, 173
  spectrum, 232, 235
  structure, 173
Concord, 47
Coniston/Adamski photo analysis, 254
control of vehicle, 243
cosmic cake, 155
  globular formation, 130
  matrix, 121, 149
  primary energy, 71
  rotary motion, 131
  strings, 104
costs, 77
counter rotating rings, 279
creative rays, 113 - 117
  coexisting, 196
Creighton, Gordon, 24
Crookes, Sir William, 140
crookes radiometer, 140, 141
  as a gravity analogy, 140
crop circles, field effects, 256, 257
cushion effect, 70

Cycle power, 309
cylinders centrifugal effect, 98
    magnetic analogy, 97 - 101
    rotating, 98
    suspension in fluid, 98
Czysz, Paul, 40

Dark matter, 151
Dechaux, Father Dominic Carmo, 178, 217, 241
deep space carrier, 283
    structural layout, 283, 284
De la Warr, George, 22
De Havilland Aircraft Co., 56
dematerialisation, 67, 186
Dempster, Derek, 11
desecration of planet, 73
descent from space trip, 320
design compromise, 76
dimensional change, 173, 194
Dirac, Paul, 106
disintegrating air, 324
diversity compromise, 75, 76
domino stunts, 267
doodle bomb, 49, 57
Doppler, J, 129
    effect, 116, 129
Dowding, Lord, 25
drastic change, 310
driving standards, 307
Dunn, J.W, 160, 197

Earth to moon journey, 286
Eddington, Sir Arthur, 89
Ehrenhaft, Felix, 219
    phenomena of rotating particles, 219
Einstein, Albert, 55, 83, 90, 91
ejection mass, 268
ejector engine, 39
electric charge, 96, 102
    field, 96
    people, 173, 200
    effect on compasses, 200
electrodynamically motivated vehicle, 238
electrogravitic motivation, 217
    lift, 239
electron, 72, 95, 96
    analogy of, 134
    formation of, 133 - 135

electron spectrum, 149
electrostatic attraction and repulsion of 95,
    space drive analogy, 234
    space drive, 277
electrostatically pulse charged, 235, 238
E.M. Syncomat as a universal drive, 289
    positioning, 281, 285
energy, free, clean, 106, 297, 310
    laws, 272
    locked in vacuum, 106
    primary, 71, 158
    transfer, 243
    zero point, 106
English Electric, 53
environment pollution of, 40
ether, 90
exhaust of the system, 257
    luminosity, 250
extraterrestrial space probe, 26
exponential curve, 65
    curve rates, 313

Fail safe simplicity, 30
falling egg, 246
    acceleration and deceleration of, 246
Faraday, Michael, 15, 71, 90
    cage, 236, 237
Fleischman and Pons fusion experiment
                                    63, 205, 213
    runaway PK effect, 206
flying ceiling limit, 308
flying VTOL homes, 320
flying saucer profile, 237
    review, 24
frequency harmonic, 235
    matching, 235
Friends of the Earth, 45
fuel consumption, 46
    fossil, 46, 73
    jumbo jet, 46
    new kind, 150

Gabrielli G,and Th.von Karman, 76
galaxies formation of, 131, 303
    dispersion of, 131
Galileo, 88
G.field effects, transparency, 239, 241, 277
    analogy of, 241

atmospheric effect, 241
 beam of, 257
 crater formed by, 252
 descending pulsed field, 257
 inverse square law effect, 245
 isolation from stress, 246
 magnetic analogy of, 252, 253
 motivated vehicle, 244
 rotary effects of, 257
 shielding, 264
 uniform acceleration, 246
 velocity gradient of, 245
gravitational transparency, 239, 241, 277
Girvan, Waveney, 11
global warming, 40
globular shape, 114
Good, Timothy, 21, 321
Gold, Thomas, 129
gravity,all pervading nature of, 83 - 85
 air jet analogy, 142
 artificial, 249
 as a deficit, 142 - 144
 atmospheric pressure analogy, 109
 body moved by, 248
 inferiority of, 85
 light beam analogy of, 146, 147
 mono-polar nature of, 145
 omni-directional nature of, 93
 repulsive force, 145
 resistance to screening, 84, 85
 winds caused by, 89
G.field motivated space ship, 277
Gray, Edwin, 333
greenhouse effect, 41
Greenpeace, 45
ground reference mode, 29
gyro solid state system, 80, 318
gyroscopic effects, 243, 282

Harmonic coupling, 331
Hawking, Prof.Stephen, 196
hazard, public, 306
helium - 3, 40
helicopter manoeuvreability by tilt, 283
 'white out', 306
Hendershot, Lester, 333
Hero's steam turbine, 303
heterodyne, 170
hieroglyphics, 25

Hit rate, 235
Hoffman, 90
horizontal thrust component, 236
housefly, hypothetical scenario, 79, 80
hovercraft, 48, 70, 71, 79, 80, 135
Hoyle, Sir Fred, 106, 130, 148
Hubble,E. 129
hydraulic analogies,97 - 100, 137 - 139
hydrogen, 39
 atom, 102, 103
Hynek, Dr.J.Alan, 156

Induced current, 136, 138
industrial conflict, 311, 312
inertia, 126
 of matter, 122 - 124
inertial resistance, 83
 forces, 241
interplanetary expansion, 313
 journey by G syncomat drive, 285
inverse square law, 121, 122, 245, 339
invisible substance, 182
invisibility, 173, 197
 analogy of 197 - 200

Jeans, Sir James, 90
jet propulsion, 267, 268
 ejector, 51 - 53
 engines, 46
 miniature version, 48
Jones, Sir Harold Soencer, 55
Jones, Richard Stanton, 61

Keely, John W, 265, 301
 motor enigma, 323
 vibratory research, 324
Kelvin, Lord, 55
Kerr magneto/electro optic effect, 262
kinetic heating, absence of 320

Laboratory levitation, 238
Laithwait, Prof. Eric, 261
 linear motor, 267
landing gear, 279
lateral thrust due to tilt, 252
Lawrence of Arabia, 70

Lederman, Leon, 43, 106, 155
legacy, 306
legislation required, 307
lenticular shape, 250
  aerodynamically more acceptable, 251
levitation, 56, 57, 173
  atmospheric effects, 184
  circular motion, 184
  dynamic nature, 184
  paranormally induced, 83, 174, 217
  passive nature, 184
lift on aircraft wings, 264
light effect on mediums, 175, 215
  negating effect on paranormal levitation
                                182, 183
  velocity, 104
living force, 177
lodestone, 83
Lodge, Sir Oliver, 72, 91, 183
Longbottom, Philip, 25
lorentz contraction, 111, 116
Lovell, Sir Bernard, 55
lunar surface vehicle, 315

Magnetic analogy, 97 - 101
  attraction, 94
  fictitious, 261
  pole, 282
  road, 265
  shielding, 83
  transparency, 178
magnus effect on rotating cylinder 97
marriage of convenience, 276
mars satellites, 69
materialisation, 67, 188
matter, direct conversion, 194, 105
  energy conversion, 104, 105
  permeability, 102
  Thermo chemical/nuclear, 104, 105
maze of tubes, 327
mental attitude of psychics, 177
  bugging, 165
Michelson/Morley experiment, 71, 91, 110, 111
microwaves, 242
M.O.D. 26, 50
modulations, 114, 115
  lack of symmetry, 250
  transference of 136
momentum, 127

Momota, Prof. Hiromu, 40
moon rocket time of, 195
  trip by syncomat vehicle, 309
Moray, T. Henry, 331
Morley, Dr. 52
motorways and highways, 305

Nano engineering, 26
Napier, D & Sons Ltd., 50
NASA, 40
national institute of fusion science, 40
  security, 59
need for gradual change, 320
negative mass, 259
nerve endings, 248
Newman, Joseph, 59, 335
Newton, Sir Isaac, 42, 90, 149
Noise Abatement Society, 45
nuclear disposal pits, 306
  missile balance, 45

Oblique orientation, 289
Occam's razor, 93, 94, 101, 145
octaves higher, 123, 149
opacity of matter, 232
  lock, 235
optimum design formula, 285
  surface area versus volume and
    strength/weight ratios, 284, 285
orbital analogy, 87, 88
orientation of vehicle, 243
orthographic projection, 254
  technical analysis, 255
oscillations by synchronisation, 331
oscillator mechanical, 226 - 229
out of body experience, 168
ozone smell associated with high voltage,

Pandora's box, 23
parabolic mirror, 310
parallel universe, 173, 197
paranormal effects, 84
  duplicated by physical means, 215
  identity between normal laws, 215
parlour game experiment, 229, 230
particles, 44, 99, 101

particles dispersion, 131
    dust, 219 - 221
past recorded, 211, 212
    energetic radiations of, 211, 212
Pears cyclopaedia, 42
pelton turbine, 300
peripheral jet, 70
perpetual motion, 47, 59, 297, 298
phase lagging or advancing, 242
    180° shift, 27
    shifting, 134
Planck, Max, 90, 103
planetary heating, 145
point source, 237
polarised light, 146, 147
politics, year 2000, 78
pollutants, 73
pollution, 72, 310
power specific, 76
    weight ratio EM syncomat unit, 272
    of the universe, 303
precognitive tendencies, 340
precognition, 158, 159
    analogy, 197
    dreams, 160
    dreams spread over time, 160 - 162
    instances, 159, 160
pressure differential, 86, 263
    differential dynamically initiated, 264
    differential static, 264
    drop, 98
Price, Robert, 26
proton electron interference, 133
    analogy of, 134
    forming modulations, 135
psychometry, 158, 166
    examples of, 166 - 168
psychokinetic effects on machinery, 210 - 212
    on people, 212
    energy, 173
    laboratory corroborative effects, 206 - 210
    physical contact effects, 209, 325, 333
    triggering, 173
pulsating bodies, 222 - 224
pulsed electric charges, 277
pulses high, 276
    latching, 235
push engine, 326
Puthoff, Hal, 106,
Pym, John, 83

Quadrupled redundancy, 30
quantum theory, (Max Planck), 103

Radiesthesia, 173, 184
    dowsing, 185
    explained by Theory of Unity, 185
    tuning, 185
    used by animals, 185
radionics, 22
ray absorbtion, 145
    diluted, 219
red shift, 129, 130, 132
    analogy, 129, 130
redundancy measures, 317
Reich, Wilhelm, 334
remote viewing, 159, 168
retro fire, 292
    pack, 293
ripple tank experiments, 225
    magnetic simulation, 225 - 7
rocket motors, small size, 278
    journey, 290
Rosetta Stone, 106
rotation of craft, 290
    of PK cylinder, 203, 204
    of polarised light, 262
Rutherford, Baron Ernest, 43

Sabotage, 11
satellite reconnaisance, 80
    traffic, 32
sceptics, authority of, 155, 156
Schnetski, Alexander, 106
Schroedinger, 90, 107
scout configuration, 278 - 280
screening, selective, 278
seawater desalination, 40
secret status curtain, 340
self sustaining, 300
    cruise speed, 309
serial time theory, 196
shock wave, 137
Sirius B gravitational field, 145, 247
    acceleration due to, 247
Slipher, V.M. 129
solar decaying orbit,   306
    natural incinerator, 306
sonoluminescence, 336

Source, 113
space craft potential, 315
space matrix, 183
  energy resevoir, 145
  deficit of, 235
  displacement, 235
  ride, 269
stability, 305
  of vehicle, 243, 318
Star Treck, 67, 193
Star Wars, 45
static electricity shielding, 83
steady state cosmological theory, 128, 129
  unification of, 128
strangeness factor, 156
strong electrical force, 134
submarine craft, 315
sunlight, focus beams of, 219
superman flight of, 184
  syncomat, 293 - 295
suppressed inventions, 323
Swift, Jonathon, 69
Syncomat EM, 269
  car example, 307
  drive, 285
  inclusive acronym, 235
  self perpetuating system, 298 - 300
  silence of, 319
  simplified anology of, 301
  space vehicle, orbits unnessary, 319
  technology, 235

Taylor, Prof. John, 157
Taylor Woodrow, 48
technological age, 75
  advance, 214
  exponential rate, 213
  social restraint, 213, 214
  society, 75
technology, new, 271
telepathy, 158, 162
  comat function of, 164
  examples of, 162 - 164
teleportation, 173, 185, 191, 196
  analogy of, 190 - 193
  for deep space journeys, 192
tenuity of matter, 194
Tesla, Nikola, 106, 242
thermodynamicists, 86

time, 115, 194, 195
  advance of, 68
  alternation of, 116
  clock, 116
  exiles from, 170
  machine 65
  march of, 354
  space, 116
  travel, 159
tornados, 303
total change, 305
trade unions inertia, 312
  example of, 313
traffic density, 315
trans-atmospheric vehicles, 291, 292
transitional ascent, 30
transportation, 305
  systems, 76
triggering mechanism, 173
Trubshaw, Brian, 47
turbogenerator, 298
turbojet cold, 347 - 353

UFOs C.E.3, 12
undercarriage, 279
underdeveloped areas, 304
unified field theory, 150, 155
unification of attraction and repulsion
        phenonema, 147, 148
uniform shared motion, 248
  railway trucks analogy, 248
unity of creation theory, 107, 110, 173
  acceleration, 127
  creative rays, 113
  frequency of, 113
  gravity, 127
  inertia, 126
  momentum, 127
  velocity, 126
  wavelength, 113
universe, 282
  overlaying, 197

Vampire, goblin, 49
vehicles, propulsion of, 76
Verschoyle, W.D., 239, 242, 259, 332
vertical thrust, 238
vibrating bodies, 226, 227

Vindication of a scout, 278
Vinci, Leonardo da, 88
visual fly up, 290
Vlitoss, Prof. Chuck, 73
volatile fuel zero hazard, 320

Waste products, 257, 259
water waves, 224
    formation of, 225
    phasing, 225, 235
    motion due to, 226
Watson, Thomas, 55
waveicle, 90, 103, 136
waves multifarious spectrum, 259
    spurious, 259
weightlessness, 32
    effect on astronauts, 181
    effect on mystics, 181
    physiological hazards, 249
    related to transparency, 244
Werner Laurie, T, 24
Wheeler, Dr. John, 149
Whittle, Sir Frank, 55, 70
Wilson, Prof. Erasmus, 55
wimshurst generator, 279
wood for the trees, 53
working fluid, 300
World War 2, 42
Wright brothers, 75, 83
Wright, Dr. John, 329

Yes, Prime Minister, 12

NEW BOOKS

THE TESLA PAPERS
Nikola Tesla on Free Energy & Wireless Transmission of Power
by Nikola Tesla, edited by David Hatcher Childress

In the tradition of *The Fantastic Inventions of Nikola Tesla*, *The Anti-Gravity Handbook* and *The Free-Energy Device Handbook*, science and UFO author David Hatcher Childress takes us into the incredible world of Nikola Tesla and his amazing inventions. Tesla's rare article "The Problem of Increasing Human Energy with Special Reference to the Harnessing of the Sun's Energy" is included. This lengthy article was originally published in the June 1900 issue of *The Century Illustrated Monthly Magazine* and it was the outline for Tesla's master blueprint for the world. Tesla's fantastic vision of the future, including wireless power, anti-gravity, free energy and highly advanced solar power.

Also included are some of the papers, patents and material collected on Tesla at the Colorado Springs Tesla Symposiums, including papers on:
•The Secret History of Wireless Transmission •Tesla and the Magnifying Transmitter
•Design and Construction of a half-wave Tesla Coil •Electrostatics: A Key to Free Energy
•Progress in Zero-Point Energy Research •Electromagnetic Energy from Antennas to Atoms
•Tesla's Particle Beam Technology •Fundamental Excitatory Modes of the Earth-Ionosphere Cavity
325 PAGES. 8X10 PAPERBACK. ILLUSTRATED. $16.95. CODE: TTP

MYSTERY IN ACAMBARO
Did Dinosaurs Survive Until Recently?
by Charles Hapgood, introduction by David Hatcher Childress

Maps of the Ancient Sea Kings author Hapgood's rare book *Mystery in Acambaro* is back in print! Hapgood researched the Acambaro collection of clay figurines with Earl Stanley Gardner (author of the Perry Mason mysteries) in the mid-1960s. The Acambaro collection comprises hundreds of clay figurines that are apparently thousands of years old; however, they depict such bizarre animals and scenes that most archaeologists dismiss them as an elaborate hoax. The collection shows humans interacting with dinosaurs and various other "monsters" such as horned men. Both Hapgood and Erle Stanley Gardner were convinced that the figurines from Acambaro were authentic ancient artifacts that indicated that men and dinosaurs had cohabited in the recent past, and that dinosaurs had not become extinct many millions of years ago as commonly thought. David Hatcher Childress writes a lengthy introduction concerning Acambaro, the latest testing, and other evidence of "living" dinosaurs.
256 PAGES. 6X9 PAPERBACK. ILLUSTRATED. BIBLIOGRAPHY. $14.95. CODE: MIA

THE BOOK OF ENOCH
The Prophet
translated by Richard Laurence

This is a reprint of the Apocryphal *Book of Enoch the Prophet* which was first discovered in Abyssinia in the year 1773 by a Scottish explorer named James Bruce. In 1821 *The Book of Enoch* was translated by Richard Laurence and published in a number of successive editions, culminating in the 1883 edition. One of the main influences from the book is its explanation of evil coming into the world with the arrival of the "fallen angels." Enoch acts as a scribe, writing up a petition on behalf of these fallen angels, or fallen ones, to be given to a higher power for ultimate judgment. Christianity adopted some ideas from Enoch, including the Final Judgment, the concept of demons, the origins of evil and the fallen angels, and the coming of a Messiah and ultimately, a Messianic kingdom. The *Book of Enoch* was ultimately removed from the Bible and banned by the early church. Copies of it were found to have survived in Ethiopia, and fragments in Greece and Italy. Like the Dead Sea Scrolls and the Nag Hammadi Library, the *Book of Enoch*, translated from the original Ethiopian Coptic script, is a rare resource that was suppressed by the early church and thought destroyed. Today it is back in print in this expanded, deluxe edition, using the original 1883 revised text.
224 PAGES. 6X9 PAPERBACK. ILLUSTRATED. INDEX. $16.95. CODE: BOE

MYSTERIES OF ANCIENT SOUTH AMERICA
Atlantis Reprint Series
by Harold T. Wilkins

The reprint of Wilkins' classic book on the megaliths and mysteries of South America. This book predates Wilkin's book *Secret Cities of Old South America* published in 1952. *Mysteries of Ancient South America* was first published in 1947 and is considered a classic book of its kind. With diagrams, photos and maps, Wilkins digs into old manuscripts and books to bring us some truly amazing stories of South America: a bizarre subterranean tunnel system; lost cities in the remote border jungles of Brazil; legends of Atlantis in South America; cataclysmic changes that shaped South America; and other strange stories from one of the world's great researchers. Chapters include: Our Earth's Greatest Disaster, Dead Cities of Ancient Brazil, The Jungle Light that Shines by Itself, The Missionary Men in Black: Forerunners of the Great Catastrophe, The Sign of the Sun: The World's Oldest Alphabet, Sign-Posts to the Shadow of Atlantis, The Atlanean "Subterraneans" of the Incas, Tiahuanacu and the Giants, more.
236 PAGES. 6X9 PAPERBACK. ILLUSTRATED. INDEX. $14.95. CODE: MASA

WAKE UP DOWN THERE!
The Excluded Middle Anthology
by Greg Bishop

The great American tradition of dropout culture makes it over the millennium mark with a collection of the best from *The Excluded Middle*, the critically acclaimed underground zine of UFOs, the paranormal, conspiracies, psychedelia, and spirit. Contributions from Robert Anton Wilson, Ivan Stang, Martin Kottmeyer, John Shirley, Scott Corrales, Adam Gorightly and Robert Sterling; and interviews with James Moseley, Karla Turner, Bill Moore, Kenn Thomas, Richard Boylan, Dean Radin, Joe McMoneagle, and the mysterious Ira Einhorn (an *Excluded Middle* exclusive). Includes full versions of interviews and extra material not found in the newsstand versions.
420 PAGES. 8X11 PAPERBACK. ILLUSTRATED. $25.00. CODE: WUDT

24 hour credit card orders—call: 815-253-6390 fax: 815-253-6300
email: auphq@frontiernet.net   www.adventuresunlimited.co.nz   www.wexclub.com

ANTI-GRAVITY

COSMIC MATRIX
Piece for a Jig-Saw, Part Two
by Leonard G. Cramp

Leonard G. Cramp, a British aerospace engineer, wrote his first book *Space Gravity and the Flying Saucer* in 1954. Cosm Matrix is the long-awaited sequel to his 1966 book *UFOs & Anti-Gravity: Piece for a Jig-Saw*. Cramp has had a long histor of examining UFO phenomena and has concluded that UFOs use the highest possible aeronautic science to move in the wa they do. Cramp examines anti-gravity effects and theorizes that this super-science used by the craft—described in detail in th book—can lift mankind into a new level of technology, transportation and understanding of the universe. The book takes close look at gravity control, time travel, and the interlocking web of energy between all planets in our solar system wi Leonard's unique technical diagrams. A fantastic voyage into the present and future!
364 PAGES. 6x9 PAPERBACK. ILLUSTRATED. BIBLIOGRAPHY. $16.00. CODE: CMX

UFOS AND ANTI-GRAVITY
Piece For A Jig-Saw
by Leonard G. Cramp

Leonard G. Cramp's 1966 classic book on flying saucer propulsion and suppressed technology is a highly technical look at the UFO phenomena by a trained scientist. Cramp first introduces the idea of 'anti-gravity' and introduces us to the various theories of gravitation. He then examines the technol- ogy necessary to build a flying saucer and examines in great detail the technical aspects of such a craft. Cramp's book is a wealth of material and diagrams on flying saucers, anti-gravity, suppressed technology, G-fields and UFOs. Chapters include Crossroads of Aerodynamics, Aerodynamic Sau- cers, Limitations of Rocketry, Gravitation and the Ether, Gravitational Spaceships, G-Field Lift Ef- fects, The Bi-Field Theory, VTOL and Hovercraft, Analysis of UFO photos, more.
388 PAGES. 6x9 PAPERBACK. ILLUSTRATED. $16.95. CODE: UAG

THE HARMONIC CONQUEST OF SPACE
by Captain Bruce Cathie

Chapters include: Mathematics of the World Grid; the Harmonics of Hiroshima and Nagasaki; Harmonic Transmission an Receiving; the Link Between Human Brain Waves; the Cavity Resonance between the Earth; the Ionosphere and Gravity Edgar Cayce—the Harmonics of the Subconscious; Stonehenge; the Harmonics of the Moon; the Pyramids of Mars; Niko Tesla's Electric Car; the Robert Adams Pulsed Electric Motor Generator; Harmonic Clues to the Unified Field; and more. Als included are tables showing the harmonic relations between the earth's magnetic field, the speed of light, and anti-gravity gravity acceleration at different points on the earth's surface. New chapters in this edition on the giant stone spheres of Cost Rica, Atomic Tests and Volcanic Activity, and a chapter on Ayers Rock analysed with Stone Mountain, Georgia.
248 PAGES. 6x9. PAPERBACK. ILLUSTRATED. BIBLIOGRAPHY. $16.95. CODE: HCS

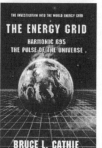

THE ENERGY GRID
Harmonic 695, The Pulse of the Universe
by Captain Bruce Cathie.

This is the breakthrough book that explores the incredible potential of the Energy Grid and the Earth's Unified Field all aroun us. Cathie's first book, *Harmonic 33*, was published in 1968 when he was a commercial pilot in New Zealand. Since then Captain Bruce Cathie has been the premier investigator into the amazing potential of the infinite energy that surrounds o planet every microsecond. Cathie investigates the Harmonics of Light and how the Energy Grid is created. In this amazin book are chapters on UFO Propulsion, Nikola Tesla, Unified Equations, the Mysterious Aerials, Pythagoras & the Grid, Nuclea Detonation and the Grid, Maps of the Ancients, an Australian Stonehenge examined, more.
255 PAGES. 6X9 TRADEPAPER. ILLUSTRATED. $15.95. CODE: TEG

THE BRIDGE TO INFINITY
Harmonic 371244
by Captain Bruce Cathie

Cathie has popularized the concept that the earth is crisscrossed by an electromagnetic grid system that can be used for anti-gravity, free energy, levitation and more. The book includes a new analysis of the harmonic nature of reality, acoustic levitation, pyramid power, harmonic receiver towers and UFO propulsion. It concludes that today's scientists have at their command a fantastic store of knowledge with which to advance the welfare of the human race.
204 PAGES. 6X9 TRADEPAPER. ILLUSTRATED. $14.95. CODE: BTF

MAN-MADE UFOS 1944—1994
Fifty Years of Suppression
by Renato Vesco & David Hatcher Childress

A comprehensive look at the early "flying saucer" technology of Nazi Germany and the genesis of man-made UFOs. This book takes us from the work of captured German scientists to escaped battalions of Germans, secre communities in South America and Antarctica to todays state-of-the-art "Dreamland" flying machines. Heavily illustrated this astonishing book blows the lid off the "government UFO conspiracy" and explains with technical diagrams the technol ogy involved. Examined in detail are secret underground airfields and factories; German secret weapons; "suction" aircraft the origin of NASA; gyroscopic stabilizers and engines; the secret Marconi aircraft factory in South America; and more Introduction by W.A. Harbinson, author of the Dell novels *GENESIS* and *REVELATION*.
318 PAGES. 6x9 PAPERBACK. ILLUSTRATED. INDEX & FOOTNOTES. $18.95. CODE: MMU

ANTI-GRAVITY

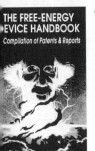

FREE ENERGY SYSTEMS

LOST SCIENCE
by Gerry Vassilatos

Rediscover the legendary names of suppressed scientific revolution—remarkable lives, astounding discoveries, and incredible inventions w[ ] would have produced a world of wonder. How did the aura research of Baron Karl von Reichenbach prove the vitalistic theory and frighten greatest minds of Germany? How did the physiophone and wireless of Antonio Meucci predate both Bell and Marconi by decades? How does earth battery technology of Nathan Stubblefield portend an unsuspected energy revolution? How did the geoaetheric engines of Nikola Te threaten the establishment of a fuel-dependent America? The microscopes and virus-destroying ray machines of Dr. Royal Rife provided the solu for every world-threatening disease. Why did the FDA and AMA together condemn this great man to Federal Prison? The static crashes on teleph lines enabled Dr. T. Henry Moray to discover the reality of radiant space energy. Was the mysterious "Swedish stone," the powerful mineral wh Dr. Moray discovered, the very first historical instance in which stellar power was recognized and secured on earth? Why did the Air Force init[ ] fund the gravitational warp research and warp-cloaking devices of T. Townsend Brown and then reject it? When the controlled fusion devices of P[ ] Farnsworth achieved the "break-even" point in 1967 the FUSOR project was abruptly cancelled by ITT.
304 PAGES. 6x9 PAPERBACK. ILLUSTRATED. BIBLIOGRAPHY. $16.95. CODE: LOS

SECRETS OF COLD WAR TECHNOLOGY
Project HAARP and Beyond
by Gerry Vassilatos

Vassilatos reveals that "Death Ray" technology has been secretly researched and developed since the turn of the century. Included are chapters such inventors and their devices as H.C. Vion, the developer of auroral energy receivers; Dr. Selim Lernstrom's pre-Tesla experiments; the ea[ ] beam weapons of Grindell-Mathews, Ulivi, Turpain and others; John Hettenger and his early beam power systems. Learn about Project Arg[ ] Project Teak and Project Orange; EMP experiments in the 60s; why the Air Force directed the construction of a huge Ionospheric "backscatt[ ] telemetry system across the Pacific just after WWII; why Raytheon has collected every patent relevant to HAARP over the past few years; mor[ ]
250 PAGES. 6x9 PAPERBACK. ILLUSTRATED. $15.95. CODE: SCWT

THE A.T. FACTOR
A Scientists Encounter with UFOs: Piece For A Jigsaw Part 3
by Leonard Cramp

British aerospace engineer Cramp began much of the scientific anti-gravity and UFO propulsion analysis back in 1955 with his landmark bo[ ] Space, Gravity & the Flying Saucer (out-of-print and rare). His next books (available from Adventures Unlimited) UFOs & Anti-Gravity: Piece[ ] a Jig-Saw and The Cosmic Matrix: Piece for a Jig-Saw Part 2 began Cramp's in depth look into gravity control, free-energy, and the interlocking w[ ] of energy that pervades the universe. In this final book, Cramp brings to a close his detailed and controversial study of UFOs and Anti-Gravity.
324 PAGES. 6x9 PAPERBACK. ILLUSTRATED. BIBLIOGRAPHY. INDEX. $16.95. CODE: ATF

THE TIME TRAVEL HANDBOOK
A Manual of Practical Teleportation & Time Travel
edited by David Hatcher Childress

In the tradition of The Anti-Gravity Handbook and The Free-Energy Device Handbook, science and UFO author David Hatcher Childress takes us into the wein world of time travel and teleportation. Not just a whacked-out look at science fiction, this book is an authoritative chronicling of real-life time travel experiment teleportation devices and more. The Time Travel Handbook takes the reader beyond the government experiments and deep into the uncharted territory of earl time travellers such as Nikola Tesla and Guglielmo Marconi and their alleged time travel experiments, as well as the Wilson Brothers of EMI and their connection to the Philadelphia Experiment—the U.S. Navy's forays into invisibility, time travel, and teleportation. Childress looks into the claims of time travellin individuals, and investigates the unusual claim that the pyramids on Mars were built in the future and sent back in time. A highly visual, large format book, wit patents, photos and schematics. Be the first on your block to build your own time travel device!
316 PAGES. 7x10 PAPERBACK. ILLUSTRATED. $16.95. CODE: TTH

THE TESLA PAPERS
Nikola Tesla on Free Energy & Wireless Transmission of Power
by Nikola Tesla, edited by David Hatcher Childress

David Hatcher Childress takes us into the incredible world of Nikola Tesla and his amazing inventions. Tesla's rare article "The Problem Increasing Human Energy with Special Reference to the Harnessing of the Sun's Energy" is included. This lengthy article was originally publish in the June 1900 issue of The Century Illustrated Monthly Magazine and it was the outline for Tesla's master blueprint for the world. Tesla's fantas vision of the future, including wireless power, anti-gravity, free energy and highly advanced solar power. Also included are some of the pape patents and material collected on Tesla at the Colorado Springs Tesla Symposiums, including papers on: •The Secret History of Wireless Transm sion •Tesla and the Magnifying Transmitter •Design and Construction of a Half-Wave Tesla Coil •Electrostatics: A Key to Free Energy •Progress Zero-Point Energy Research •Electromagnetic Energy from Antennas to Atoms •Tesla's Particle Beam Technology •Fundamental Excitato Modes of the Earth-Ionosphere Cavity
325 PAGES. 8x10 PAPERBACK. ILLUSTRATED. $16.95. CODE: TTP

THE FANTASTIC INVENTIONS OF NIKOLA TESLA
by Nikola Tesla with additional material by David Hatcher Childress

This book is a readable compendium of patents, diagrams, photos and explanations of the many incredible inventions of the originator of the moder era of electrification. In Tesla's own words are such topics as wireless transmission of power, death rays, and radio-controlled airships. In additio rare material on German bases in Antarctica and South America, and a secret city built at a remote jungle site in South America by one of Tesla students, Guglielmo Marconi. Marconi's secret group claims to have built flying saucers in the 1940s and to have gone to Mars in the early 1950[ ] Incredible photos of these Tesla craft are included. The Ancient Atlantean system of broadcasting energy through a grid system of obelisks and pyramids is discussed, and a fascinating concept comes out of one chapter: that Egyptian engineers had to wear protective metal head-shields whil in these power plants, hence the Egyptian Pharoah's head covering as well as the Face on Mars! •His plan to transmit free electricity into the atmosphere. •How electrical devices would work using only small antennas. •Why unlimited power could be utilized anywhere on earth. •Ho[ ] radio and radar technology can be used as death-ray weapons in Star Wars.
342 PAGES. 6x9 PAPERBACK. ILLUSTRATED. $16.95. CODE: FINT

24 hour credit card orders—call: 815-253-6390 fax: 815-253-6300
email: auphq@frontiernet.net   www.adventuresunlimitedpress.com   www.wexclub.com

STRANGE SCIENCE

UNDERGROUND BASES & TUNNELS
What is the Government Trying to Hide?
by Richard Sauder, Ph.D.
Working from government documents and corporate records, Sauder has compiled an impressive book that digs below the surface of the military's super-secret underground! Go behind the scenes into little-known corners of the public record and discover how corporate America has worked hand-in-glove with the Pentagon for decades, dreaming about, planning, and actually constructing, secret underground bases. This book includes chapters on the locations of the bases, the tunneling technology, various military designs for underground bases, nuclear testing & underground bases, abductions, needles & implants, military involvement in "alien" cattle mutilations, more. 50 page photo & map insert.
201 PAGES. 6X9 PAPERBACK. ILLUSTRATED. $15.95. CODE: UGB

KUNDALINI TALES
by Richard Sauder, Ph.D.
Underground Bases and Tunnels author Richard Sauder's second book on his personal experiences and provocative research into spontaneous spiritual awakening, out-of-body journeys, encounters with secretive governmental powers, daylight sightings of UFOs, and more. Sauder continues his studies of underground bases with new information on the occult underpinnings of the U.S. space program. The book also contains a breakthrough section that examines actual U.S. patents for devices that manipulate minds and thoughts from a remote distance. Included are chapters on the secret space program and a 130-page appendix of patents and schematic diagrams of secret technology and mind control devices.
296 PAGES. 7X10 PAPERBACK. ILLUSTRATED. BIBLIOGRAPHY. $14.95. CODE: KTAL

LOST SCIENCE
by Gerry Vassilatos
Secrets of Cold War Technology author Vassilatos on the remarkable lives, astounding discoveries, and incredible inventions of such famous people as Nikola Tesla, Dr. Royal Rife, T.T. Brown, and T. Henry Moray. Read about the aura research of Baron Karl von Reichenbach, the wireless technology of Antonio Meucci, the controlled fusion devices of Philo Farnsworth, the earth battery of Nathan Stubblefield, and more. What were the twisted intrigues which surrounded the often deliberate attempts to stop this technology? Vassilatos claims that we are living hundreds of years behind our intended level of technology and we must recapture this "lost science."
304 PAGES. 6X9 PAPERBACK. ILLUSTRATED. BIBLIOGRAPHY. $16.95. CODE: LOS

THE TIME TRAVEL HANDBOOK
A Manual of Practical Teleportation & Time Travel
edited by David Hatcher Childress
In the tradition of *The Anti-Gravity Handbook* and *The Free-Energy Device Handbook*, science and UFO author David Hatcher Childress takes us into the weird world of time travel and teleportation. Not just a whacked-out look at science fiction, this book is an authoritative chronicling of real-life time travel experiments, teleportation devices and more. *The Time Travel Handbook* takes the reader beyond the government experiments and deep into the uncharted territory of early time travellers such as Nikola Tesla and Guglielmo Marconi and their alleged time travel experiments, as well as the Wilson Brothers of EMI and their connection to the Philadelphia Experiment—the U.S. Navy's forays into invisibility, time travel, and teleportation. Childress looks into the claims of time travelling individuals, and investigates the unusual claim that the pyramids on Mars were built in the future and sent back in time. A highly visual, large format book, with patents, photos and schematics. Be the first on your block to build your own time travel device!
316 PAGES. 7X10 PAPERBACK. ILLUSTRATED. $16.95. CODE: TTH

MAPS OF THE ANCIENT SEA KINGS
Evidence of Advanced Civilization in the Ice Age
by Charles H. Hapgood
Charles Hapgood's classic 1966 book on ancient maps produces concrete evidence of an advanced world-wide civilization existing many thousands of years before ancient Egypt. He has found the evidence in the Piri Reis Map that shows Antarctica, the Hadji Ahmed map, the Oronteus Finaeus and other amazing maps. Hapgood concluded that these maps were made from more ancient maps from the various ancient archives around the world, now lost. Not only were these unknown people more advanced in mapmaking than any people prior to the 18th century, it appears they mapped all the continents. The Americas were mapped thousands of years before Columbus. Antarctica was mapped when its coasts were free of ice.
316 PAGES. 7X10 PAPERBACK. ILLUSTRATED. BIBLIOGRAPHY & INDEX. $19.95. CODE: MASK

PATH OF THE POLE
Cataclysmic Pole Shift Geology
by Charles Hapgood
Maps of the Ancient Sea Kings author Hapgood's classic book *Path of the Pole* is back in print! Hapgood researched Antarctica, ancient maps and the geological record to conclude that the Earth's crust has slipped in the inner core many times in the past, changing the position of the pole. *Path of the Pole* discusses the various "pole shifts" in Earth's past, giving evidence for each one, and moves on to possible future pole shifts. Packed with illustrations, this is the sourcebook for many other books on cataclysms and pole shifts.
356 PAGES. 6X9 PAPERBACK. ILLUSTRATED. $16.95. CODE: POP.

THE LOST CITIES SERIES

LOST CITIES OF ATLANTIS, ANCIENT EUROPE & THE MEDITERRANEAN
by David Hatcher Childress
Atlantis! The legendary lost continent comes under the close scrutiny of maverick archaeologist David Hatcher Childress in sixth book in the internationally popular *Lost Cities* series. Childress takes the reader in search of sunken cities in the Medite nean; across the Atlas Mountains in search of Atlantean ruins; to remote islands in search of megalithic ruins; to meet liv legends and secret societies. From Ireland to Turkey, Morocco to Eastern Europe, and around the remote islands of the Med ranean and Atlantic, Childress takes the reader on an astonishing quest for mankind's past. Ancient technology, cataclys megalithic construction, lost civilizations and devastating wars of the past are all explored in this book. Childress challenges skeptics and proves that great civilizations not only existed in the past, but the modern world and its problems are reflections o ancient world of Atlantis.
524 PAGES. 6X9 PAPERBACK. ILLUSTRATED WITH 100S OF MAPS, PHOTOS AND DIAGRAMS. BIB OGRAPHY & INDEX. $16.95. CODE: MED

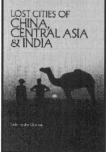

LOST CITIES OF CHINA, CENTRAL INDIA & ASIA
by David Hatcher Childress
Like a real life "Indiana Jones," maverick archaeologist David Childress takes the reader on an incredible adventure ac some of the world's oldest and most remote countries in search of lost cities and ancient mysteries. Discover ancient citie the Gobi Desert; hear fantastic tales of lost continents, vanished civilizations and secret societies bent on ruling the world; forgotten monasteries in forbidding snow-capped mountains with strange tunnels to mysterious subterranean cities! A uni combination of far-out exploration and practical travel advice, it will astound and delight the experienced traveler or armchair voyager.
429 PAGES. 6X9. PAPERBACK. PHOTOS, MAPS, AND ILLUSTRATIONS WITH FOOTNOTES & BIB OGRAPHY $14.95. CODE CHI

LOST CITIES OF ANCIENT LEMURIA & THE PACIFIC
by David Hatcher Childress
Was there once a continent in the Pacific? Called Lemuria or Pacifica by geologists, Mu or Pan by the mystics, there is ample mythological, geological and archaeological evidence to "prove" that an advanced and ancient civilization once li in the central Pacific. Maverick archaeologist and explorer David Hatcher Childress combs the Indian Ocean, Australia the Pacific in search of the surprising truth about mankind's past. Contains photos of the underwater city on Pohn explanations on how the statues were levitated around Easter Island in a clock-wise vortex movement; tales of disappea islands; Egyptians in Australia; and more.
379 PAGES. 6X9. PAPERBACK. ILLUSTRATED. FOOTNOTES & BIBLIOGRAPHY $14.95. CODE LE

LOST CITIES OF NORTH & CENTRAL AMERICA
by David Hatcher Childress
Down the back roads from coast to coast, maverick archaeologist and adventurer David Hatcher Childress goes deep unknown America. With this incredible book, you will search for lost Mayan cities and books of gold, discover an anc canal system in Arizona, climb gigantic pyramids in the Midwest, explore megalithic monuments in New England, and the astonishing quest for the lost cities throughout North America. From the war-torn jungles of Guatemala, Nicaragua Honduras to the deserts, mountains and fields of Mexico, Canada, and the U.S.A., Childress takes the reader in searc sunken ruins, Viking forts, strange tunnel systems, living dinosaurs, early Chinese explorers, and fantastic lost treas Packed with both early and current maps, photos and illustrations.
590 PAGES. 6X9 PAPERBACK. PHOTOS, MAPS, AND ILLUSTRATIONS. FOOTNOTES & BIB OGRAPHY. $14.95. CODE: NCA

LOST CITIES & ANCIENT MYSTERIES OF AFRICA & ARABIA
by David Hatcher Childress
Across ancient deserts, dusty plains and steaming jungles, maverick archaeologist David Childress continues his world-w quest for lost cities and ancient mysteries. Join him as he discovers forbidden cities in the Empty Quarter of Arabia; "Atlante ruins in Egypt and the Kalahari desert; a mysterious, ancient empire in the Sahara; and more. This is the tale of an extrac nary life on the road: across war-torn countries, Childress searches for King Solomon's Mines, living dinosaurs, the Ar the Covenant and the solutions to some of the fantastic mysteries of the past.
423 PAGES. 6X9 PAPERBACK. PHOTOS, MAPS, AND ILLUSTRATIONS. FOOTNOTES & BIBLIOGRAP $14.95. CODE: AFA

LOST CITIES & ANCIENT MYSTERIES OF SOUTH AMERICA
by David Hatcher Childress
Rogue adventurer and maverick archaeologist David Hatcher Childress takes the reader on unforgettable journeys deep i deadly jungles, high up on windswept mountains and across scorching deserts in search of lost civilizations and anci mysteries. Travel with David and explore stone cities high in mountain forests and hear fantastic tales of Inca treasu living dinosaurs, and a mysterious tunnel system. Whether he is hopping freight trains, searching for secret cities, or j dealing with the daily problems of food, money, and romance, the author keeps the reader spellbound. Includes both ea and current maps, photos, and illustrations, and plenty of advice for the explorer planning his or her own journey of disc ery.
381 PAGES. 6X9 PAPERBACK. PHOTOS, MAPS, AND ILLUSTRATIONS. FOOTNOTES & BIBLIOGR PHY. $14.95. CODE: SAM

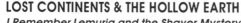

One Adventure Place
P.O. Box 74
Kempton, Illinois 60946
United States of America
Tel.: 815-253-6390 • Fax: 815-253-6300
Email: auphq@frontiernet.net
http://www.adventuresunlimitedpress.com
or www.wexclub.com/aup

ORDERING INSTRUCTIONS

✓ Remit by USD$ Check, Money Order or Credit Card
✓ Visa, Master Card, Discover & AmEx Accepted
✓ Prices May Change Without Notice
✓ 10% Discount for 3 or more Items

SHIPPING CHARGES

United States

✓ Postal Book Rate { $3.00 First Item
50¢ Each Additional Item

✓ Priority Mail { $4.00 First Item
$2.00 Each Additional Item

✓ UPS { $5.00 First Item
$1.50 Each Additional Item

NOTE: UPS Delivery Available to Mainland USA Only

Canada

✓ Postal Book Rate { $4.00 First Item
$1.00 Each Additional Item

✓ Postal Air Mail { $6.00 First Item
$2.00 Each Additional Item

✓ Personal Checks or Bank Drafts MUST BE
USD$ and Drawn on a US Bank
✓ Canadian Postal Money Orders OK
✓ Payment MUST BE USD$

All Other Countries

✓ Surface Delivery { $7.00 First Item
$2.00 Each Additional Item

✓ Postal Air Mail { $13.00 First Item
$8.00 Each Additional Item

✓ Payment MUST BE USD$
✓ Checks and Money Orders MUST BE USD$
and Drawn on a US Bank or branch.
✓ Add $5.00 for Air Mail Subscription to
Future *Adventures Unlimited* Catalogs

SPECIAL NOTES

✓ RETAILERS: Standard Discounts Available
✓ BACKORDERS: We Backorder all Out-of-
Stock Items Unless Otherwise Requested
✓ PRO FORMA INVOICES: Available on Request
✓ VIDEOS: NTSC Mode Only. Replacement only.
✓ For PAL mode videos contact our other offices:

European Office:
Adventures Unlimited, Panewaal 22,
Enkhuizen, 1600 AA, The Netherlands
http: www.adventuresunlimited.nl
Check Us Out Online at:
www.adventuresunlimitedpress.com

Please check: ☑

☐ This is my first order ☐ I have ordered before ☐ This is a new address

Name	
Address	
City	
State/Province	Postal Code
Country	
Phone day	Evening
Fax	

Item Code	Item Description	Price	Qty	Total

Please check: ☑

☐ Postal-Surface
☐ Postal-Air Mail
(Priority in USA)
☐ UPS
(Mainland USA only)

Subtotal ➙	
Less Discount-10% for 3 or more items ➙	
Balance ➙	
Illinois Residents 6.25% Sales Tax ➙	
Previous Credit ➙	
Shipping ➙	
Total (check/MO in USD$ only) ➙	

☐ Visa/MasterCard/Discover/Amex

Card Number

Expiration Date

10% Discount When You Order 3 or More Items!

Comments & Suggestions	Share Our Catalog with a Friend